ALSO BY ZACHARY KARABELL

*What's College For?*
*The Struggle to Define American Higher Education*

*Architects of Invention: The United States, the Third World,*
*and the Cold War, 1946–1962*

*The Last Campaign:*
*How Harry Truman Won the 1948 Election*

*A Visionary Nation:*
*Four Centuries of American Dreams and What Lies Ahead*

# Parting the Desert

# PARTING THE DESERT

*The Creation of the Suez Canal*

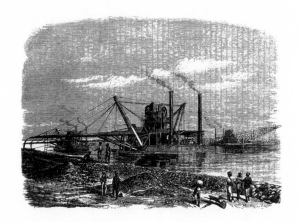

# ZACHARY KARABELL

ALFRED A. KNOPF

NEW YORK   2003

THIS IS A BORZOI BOOK
PUBLISHED BY ALFRED A. KNOPF

Copyright © 2003 by Zachary Karabell
All rights reserved under International and Pan-American Copyright
Conventions. Published in the United States by Alfred A. Knopf,
a division of Random House, Inc., New York, and simultaneously in
Canada by Random House of Canada Limited, Toronto.
Distributed by Random House, Inc., New York.
www.aaknopf.com

Knopf, Borzoi Books, and the colophon are registered
trademarks of Random House, Inc.

Library of Congress Cataloging-in-Publication Data
Karabell, Zachary.
Parting the desert : the creation of the Suez Canal / Zachary Karabell. — 1st ed.
p.    cm.
Includes bibliographical references and index.
ISBN 0-375-40883-5 (alk. paper)
1. Suez Canal (Egypt)—History.
2. Canals—Egypt—Suez—Design and construction.
3. Lesseps, Ferdinand de, 1805–1894. I. Title.

TC791 .K37 2003
386'.43'09—dc21        2002034209

Manufactured in the United States of America
First Edition

# Contents

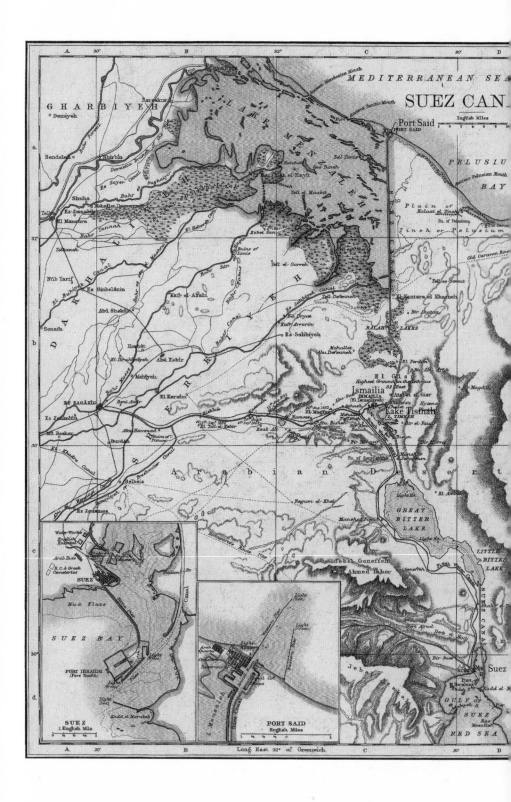

# PARTING THE DESERT

# THE TWILIGHT

I T WAS LATE afternoon in the desert when they emerged from the labyrinth of eddies that flowed through the Nile Delta. The breeze carried the trace of salt from the Mediterranean, and though the wind was less harsh than it had been several weeks before, the area was still desolate. The day had been warm; the night promised bitter cold, but at least they had finally arrived at their destination on the coast, where the easternmost branch of the Nile used to meet the Mediterranean Sea. Pitching their tents, Ferdinand de Lesseps and his companions settled in for the evening.

They had sailed from Damietta, where St. Louis died centuries before in a fool's errand of a crusade, and where much later Napoleon's troops stumbled on the Rosetta Stone. They had crossed Lake Manzala, in four fishing boats outfitted with small cabins to shelter them at night. And then they camped on the thin littoral separating the brackish lake from the waters of the Mediterranean. They were not there to explore. They were there to begin.

At dawn on April 25, 1859, they packed their camels and hurried to their destination, where they were joined by a group of Egyptian laborers. There were 150 altogether, diplomats, businessmen, engineers, and peasants. Their silhouettes moved across the sunrise, and over each shoulder there was a pickax.

At a spot known only to him, Lesseps raised his hand and ordered the company to halt. They unloaded their gear and stood, picks in hand, waiting. Lesseps looked toward the sea and then back toward the desert. His compact energy conveyed a surprising vigor for a fifty-three-year-old widowed former bureaucrat. His eyes set, his mustache elegantly trimmed, he was at a pivotal point in his life, and he knew it.

3

It was an act of theater, carefully staged. No official representatives of any government attended the ceremony, and the story of what happened was disseminated only by Lesseps himself. He unfurled an Egyptian flag and planted it in the ground, yet there was something furtive about the whole endeavor. He had sought but not formally received the blessing of the ruler of Egypt. He decided to proceed anyway. He paused for a moment, not just because he was taking a risk, but because he was about to change the political landscape of three continents, because he was embarking on an adventure that would alter the terrain of the planet. Then he spoke.

"In the name of the Universal Company of the Maritime Suez Canal, we are about to commence this work, which will open up the East to the commerce and civilization of the West. . . . The thorough surveys that we have done give us the confidence that the enterprise that commences today will not only be a work of progress, but will return immense rewards to those who have striven to make it real." He told the group to lift up their axes. "Remember," he continued, "you are not simply digging up soil. Your work will bring prosperity to your families and to your countries." Then he asked one of the workers to hand him an ax. Shouting, "In honor of the viceroy Muhammad Said Pasha," he raised his arm, and 150 arms raised up with him. Row after row of metal picks gleamed in the sun and descended into the earth. The building of the Suez Canal had begun.[1]

The states of Europe competed over it; the Ottoman Empire tried to prevent its construction; and, later, the armies of the modern Middle East destroyed the cities along its banks. In 1869, the kings and queens of Europe gathered to celebrate its inauguration, for a week of festivities so lavish that even the jaded royalty of Paris, London, and Vienna were awed. It was *the* triumph of the mid-nineteenth century, a joint venture between the ruler of Egypt and an ambitious Frenchman. Heralded as a symbol of progress, lauded as proof that geography would no longer separate the Orient and the Occident, the East and the West, the Suez Canal was the center of the world.

Today, all that marks its southern point is a pediment encased in graffiti. No statue, not even a memory of the forgotten soldier who once stood there to commemorate Egypt's wars. A few rusting parts of tanks serve as a reminder of the battles fought between Egypt and Israel, but that seems long ago. The day when Gamal Abdel Nasser stood in a square in Alexandria in 1956 and defied the French, the English, and the Americans by nationalizing the canal—even that belongs to another era, when Suez still mattered, when its fate could send the world into panic.

Where fleets once sailed, now there are tankers from third-rank nations. Freighters from Doha and container ships from Monrovia glide past that ruined pediment at the southern tip. They enter or exit the canal in convoy clusters of four or five. On the journey south, they spill out into the azure waters of the Gulf of Suez, where at sunset, the sharp brown cliffs of the Gebel Attaka rise quickly to block the dying rays.

On the journey north, they make the hundred-mile voyage to the Mediterranean Sea in less than a day. They sail through the Bitter Lakes and past the city of Ismailia, until they empty into Lake Manzala and arrive at the terminus of Port Said. There they pass by the head-quarters of the Canal Company, a decaying Victorian fantasy of white crenellation and green domes. Nearby is another orphaned stone pedestal, where a statue of Ferdinand de Lesseps once stood. Freighters and tankers pass through a canal bounded not by heroic monuments, but by pockmarked pediments.

The canal was built by the sweat of hundreds of thousands, but it was Lesseps's child—his, and that of the rulers of Egypt, Said and Ismail. Lesseps, a French diplomat who worshiped at the altar of progress; Muhammad Said, the Egyptian viceroy who saw in France and England all that his country could be but wasn't; and Ismail, his successor, who tried to purchase a better future for Egypt using European loans. Lesseps saw his vision vindicated. Said and Ismail died before realizing that theirs would not be. But decades later, enraged at the lost promise of the canal, their great-grandchildren seized it for themselves.

In July 1956, Egyptians listened as Nasser claimed the canal for Egypt and believed that after years of disappointment, the future had finally arrived. They thought that the progress promised in the nineteenth century had only been delayed until the twentieth. But their moment of triumph was brief. The British, French, and Israelis attacked, and the city of Port Said was partially destroyed by British jets. Nasser scuttled ships in the canal to prevent its use by Egypt's enemies. A decade later, during the 1967 Arab-Israeli war, all of the cities along the canal were bombarded, this time by Israeli forces on the eastern bank. Ships were scuttled once again; the harbors were mined; and the canal was closed for years. A war of attrition followed, and Ismailia was decimated. In 1973, Suez was a battle zone one more time, when Anwar Sadat ordered his army to cross the canal and confront Israeli troops on the other side.

Today, half a million people live in the reconstructed cities of Suez and Ismailia, and a million in Port Said. The port of Suez has been inhabited since Roman times, but today it is a concrete jungle of prefab apartment blocks, with no trace of anything earlier than the 1970s. Ismailia nestles Lake Timsah, where Moses may or may not have led the people of Israel out of Egypt thirty-five centuries ago. Its shores are now guarded with barbed wire. A few Victorian houses survived the Arab-Israeli wars. The mansion where Lesseps and his family lived still stands. It is a museum filled with bad copies of nineteenth-century furniture and Parisian drapes, closed to the public except on permission from the Suez Canal authorities, which must be obtained in Cairo many days in advance. Port Said, the jewel of the canal, which guards its Mediterranean entrance, is cloaked in pollution. Miles of anonymous buildings give way to petroleum-soaked beaches that seasonally fill with middle-class Egyptians on holiday. The main street is one of the prime shopping centers in the country, but only for a few months each year, when those thousands of vacationers arrive to lie on the shore and watch the ships as they line up to make the journey south.

The canal still makes money for Egypt, but the commerce of the world has found other outlets. When the canal was completed, the journey from Europe to India was sliced from months to weeks, and the arduous route around the Cape of Good Hope was rendered obsolete. The twentieth century reversed the process. Modern tankers and container ships can take the longer route, and the cost is not much different. The trip around the Cape requires more fuel, but the shorter route entails canal dues, and for many companies, it is six of one, half a dozen of the other. Some newer ships are too large for the canal, and the Egyptian government is left with a Sisyphean task: by the time it dredges and widens the channel, the next generation of supertankers will have arrived. The navies of the world take advantage of the canal for reasons of convenience, but outside of Egypt or Israel, Jordan or Saudi Arabia, it is no one's strategic priority.

That was not the future its creators imagined. A vision of progress energized Lesseps, a vision that East and West could be joined, and that the union of the two seas and the two worlds would allow the energies of mankind to flourish as never before. A similar dream animated the rulers of Egypt, the emperor and empress of France, the engineers who designed the canal, and the shareholders who invested in it. In the middle of the nineteenth century, mankind was on the verge of conquering the world. Nature would be harnessed for the betterment of all. Disease and ignorance would be no more. Where superstition and archaic religion had once kept people fearful and timid, science and industry would allow human beings to claim the birthright of Eden.

Ferdinand de Lesseps was a potent combination of vision, pragmatism, and will. Without vision, Lesseps could never have turned a hundred-mile stretch of barely inhabited desert with no source of fresh water into a canal connecting the Mediterranean to the Red Sea. He could not have mobilized the leaders of Egypt, the financial elites of Europe, Napoleon III, and millions of French citizens to support his improbable scheme. He could never have dodged the opposition of the British government and its wily prime minister, Lord Palmerston, or the animosity of the Ottoman sultan and his ministers in Constantinople. Without pragmatism, Lesseps could never have organized a venture that required the largest jetty ever constructed and a decade of labor by tens of thousands of workers. He could not have convinced engineers to design machines to dredge the soil; he could not have set up a company to oversee the excavation of seventy-five million cubic meters of sand; he could not have maneuvered through the diplomatic shoals that constantly loomed. And without will, he could not have fended off others who wanted to claim the idea for themselves.

He was not alone, however. In antiquity, there had been canals linking the Red Sea to the Mediterranean—not direct ones, for the most part, but canals nonetheless, using the Nile to Cairo and a waterway from Cairo to the port of Suez. And when those fell into disuse and silted over, the memory of them remained, kept alive by travelers' tales and by European merchants looking for alternate sea routes that would make them rich and their nations strong.

The idea for the modern canal was born before Lesseps, at a decisive moment in modern Egyptian history, when Napoleon Bonaparte landed outside of Alexandria in 1798. He brought with him not just an army, but a team of scholars charged with studying the country so that the French could make maximal use of its resources. They knew of the old canals and hoped to dig another one, but an erroneous calculation by the leading surveyor set back plans for decades, until a group of engineers, led by a disciple of Henri de Saint-Simon, returned to Egypt thirty-five years later.

If Napoleon reintroduced the idea of a maritime canal, Prosper Enfantin and the Saint-Simonians took it one step closer to fruition. They believed in progress, but even more, they believed that the world had been sundered. Male stood apart from female, religion stood opposed to science, and East was cut off from West. In this duality, said Enfantin, all suffered. The West was advancing, but until it was united with the East, its growth would be stunted. The only way to heal these wounds was to join the two worlds by removing the physical obstacle that separated them, the land bridge of the Isthmus of Suez. Once the canal pierced the sands and the seas were connected, the energies of East and West would flow together. The world would be made whole, and civilization would blossom.

Whatever the virtues of Enfantin's scheme, its metaphysical trappings limited its appeal. He and his followers barely escaped prison on their journey to Egypt, and had it not been for the intervention of the French consul, they might not have fared well in the land of the pharaohs. But although young Ferdinand de Lesseps protected the Saint-Simonians in the 1830s, twenty years later they came to regret his help. Having saved them, he then appropriated their ambition and made it his own. Lesseps was not the first person to seize someone else's idea, but he was surely one of the most successful at it. In later years, his name was immortalized. He was indelibly associated with the canal, and the contribution of Enfantin and the Saint-Simonians was largely forgotten.

But it took more than Lesseps's imagination and the industry of Europe. The Saint-Simonians dreamed of a marriage between East and West, but no matter how eager the West was for this union, without the active assent of the East the marriage could never have been consummated. To make the canal a reality, Lesseps needed the support of the ruler of Egypt, and he found a willing partner in Muhammad Said Pasha.

Said was the son of Muhammad Ali, an Albanian mercenary who rose to power in Egypt in the wake of Napoleon's defeat. He was a fat child, fond of food and disinclined to the manly virtues of exercise and martial prowess. His father was determined to make Said into a man who could practice the arts of war and diplomacy with equal skill, and he entrusted Said to the care of the French consul. The young prince and Lesseps forged a bond, and when Said came to power in 1854, he invited Lesseps to celebrate his new fortune. Hoping to convince Said of the virtues of the canal, Lesseps happily accepted the invitation.

Though later generations of Europeans and Egyptians dismissed Said as a corpulent dilettante, he had a dream no less expansive than Lesseps's, and no less improbable than Enfantin's. He was intoxicated by the promise of an Egypt restored to prominence, no longer under the control of the Ottoman Empire, and once again thriving on the valuable resources of the Nile Valley and the trade of the eastern Mediterranean. Said's vision converged with that of Lesseps, and an improbable scheme was transformed into a project that would drain the resources of Egypt, strain the balance of power in Europe, and change the face of the earth.

The list begins with Lesseps, Enfantin, and Said, but there were others: Ismail, Said's successor, who declared that he was more committed to the canal than anyone could ever be, and without whom the work might still not have been completed; the Emperor Napoleon III and his wife, Eugénie, who set the tone for the twenty-year fantasy known as the Second Empire; the French shareholders who backed the project; the engineers who designed it; the laborers who built it; and the legions in England who defied Her Majesty's Government and its prime minister, Lord Palmerston, and rallied behind the vision of a world made whole.

East and West, Orient and Occident, were terms of imagination as much as they were geographical designations. Yet, in the nineteenth century, the elites of Europe and the Middle East took those categories seriously. Frenchmen such as Lesseps believed that they were part of a civilization called "the West," and rulers such as Said perceived themselves as part of another, older civilization called "the East." Those who governed the Ottoman Empire and Egypt in the nineteenth century understood that the West now had the power, and they recognized, often resentfully, that unless they did something drastic they would be overwhelmed. Their ambitions for a better future depended on adopting the tools of the West to defend themselves against it. They had to learn to fight by different rules and with different weapons, and that meant reorganizing their societies.

For a time, European and Middle Eastern visions of progress converged, and one of the progeny of that union was the Suez Canal. The canal was not just a monumental act of engineering and organization. It was the culmination of ideals and ambitions, and a symbol of all that the culture of the nineteenth century held dear. It was a hundred-mile-long trench that signaled the triumph of science, the creativity of mankind, and the beginning of a wonderful future.

Or so its creators thought. The canal's inauguration in 1869 was supposed to herald better days, but for Egypt that did not happen. By the grace of Suez, French shareholders and English politicians became rich, powerful, and feared. Situated at the center of the British Empire, the Suez Canal became an excuse for imperial expansion, and then a cause of imperial overstretch. But the dreams of the Egyptians that they would again rule the eastern Mediterranean and reclaim the grandeur of the pharaohs, the glory of the Ptolemies, the power of Saladin, and the vigor of the Mamelukes—those were mirages that tantalized and then evaporated. For Egypt and other states east of Suez, the day the canal opened was not the culmination of a dream but the death of one. While Europe became more dominant in the decades ahead, the Near East and much of the world struggled to maintain a modicum of independence in the face of European expansion. The twentieth century offered more of the same, and Nasser's moment of glory in 1956 was a false dawn for Egypt, for the Middle East, and for the canal.

The history of East and West seemed to be running on parallel courses in the mid-nineteenth century. Organized religion was in retreat in Europe, and in Egypt and the Middle East, few elites spoke of Islam. Europe had the edge, but individuals like Lesseps believed that, in time, the spread of Western civilization would improve the lives of everyone, everywhere. That creed was taken up by the United States in the twentieth century, but its legacy outside of Europe and the United States is ambiguous. Today, the West continues to grow rich, while the rest continue to play catch-up. Disillusioned, some in the Middle East turn to Islam to do for them what the ideals of the nineteenth century did not: to restore lost pride and past glory, and to improve the quality of life in an overpopulated society. And though many shun radical Islam, few have much appetite for Western ideals. Having placed their hope in those in the nineteenth century, they are reluctant to embrace them in the twenty-first. The Suez Canal today is a reminder of the rift between East and West, and not, as its creators had wished, a symbol of unity.

These are simply historical facts, and ones that would have leavened the enthusiasm of those who came to celebrate the canal in 1869. And had they all been transported, on some magic carpet, to the present, to sit by the banks of the modern canal, they might have wondered whether their vision had been worth the cost. Even had they known, they might have pressed ahead anyway. It seemed the right thing to do at the time, and, after all, it is hard to avoid the law of unexpected consequences, just as no one can determine history's flow. The Suez Canal is now in its twilight, but it was, for a time, an unequivocal triumph of human will and ingenuity. It is fitting, therefore, that the story of its creation begins with a young, willful general looking for a battlefield, a man of revolutionary France, born in Corsica and named Buonaparte.

# CHAPTER TWO

# THE FRENCH FALL IN LOVE

T HE AGE OF Enlightenment was also a continuation of the age of exploration. The world was shrinking, at least for the inhabitants of Western Europe, and ambitions were no longer confined to the continent. In 1672, Gottfried Wilhem von Leibniz had an idea, and he was so enamored of it that he wrote to the Sun King himself, Louis XIV, lord of France.

Leibniz was twenty-six years old and living in Nuremberg, a cosmopolitan city surrounded by warring German principalities. Though he would later become one of history's most important philosopher-mathematicians, at the time he was a minor member of the court of a German prince, and he looked at the growing power of the Sun King with considerable alarm. Hoping to deflect the energies of French expansion away from the German states, Leibniz came up with a bold scheme: he would convince Louis XIV to send an army a thousand miles away, to the shores of the southeastern corner of the Mediterranean Sea, to the land of the pharaohs, Egypt.

"This project," Leibniz wrote, "is the most grand of enterprises, and the easiest to do, exempt from peril even should it fail." It would, he continued, "procure for France the suzerainty of the seas and of commerce, and it will not require any resources other than those that have already been prepared. It will obtain for the king universal affection and dissipate all old animosities and distrust. It will make him the arbiter of Christianity. It will open the road to posterity, and for the king himself, the opportunity to be compared to Alexander the Great." It was only necessary, he said, for the king to seize his destiny.

Leibniz detailed how easy it would be for a well-organized French invasion to overcome the Mameluke dynasty, then in control of Egypt. He argued that the conquest of the country would vindicate the failed crusade of Louis IX, who had tried to reclaim the Holy Land by invading Egypt in the thirteenth century and then died of fever in Damietta. He assured the Sun King that a new expedition would allow France to exploit the agricultural riches of the Nile, and to extend the hand of the Catholic monarch to protect the Holy Sepulcher in Jerusalem. And, in passing, he mentioned the one thing that would later bear fruit, long after both he and Louis had died.

"There is in Egypt," Leibniz wrote, "the most important isthmus in the world, that separating its great seas, the Ocean and the Mediterranean: a place that cannot be avoided without circling all the sinuosities of Africa; the connecting point, the obstacle, the key, the only possible door between two areas of the world, Asia and Africa; the meeting-point and marketing-place of India on one side, and Europe on the other." Louis XIV dismissed the idea, and Leibniz, who would later invent integral calculus along with Newton, was brushed off with barely an acknowledgment from a Versailles courtier. But he had tossed a pebble into a pond and, 125 years later, its ripples reached the shores of Corsica.[1]

It had not been Napoleon's life ambition to conquer the Nile Valley, but in the early months of 1798, it was his best course of action. Having emerged victorious from a brutal campaign in Italy, he was a hero to a young French republic that desperately needed one. But although he was wildly popular with his troops and with the crowds of Paris, the oligarchy that clung to power under the name of the Directory eyed him warily. While Napoleon was troubled by unfulfilled ambition, the Directory was troubled by an unanswered question: what should be done with Napoleon, the fiery general with the newly altered surname, Bonaparte?

His preference was to invade England, which was the one power that still threatened to turn back the French Revolution. But when he reviewed the logistics of mounting a cross-channel campaign against the English, he found, as had so many before and so many after, that the risks were too great and the rewards too uncertain. That left Napoleon with an army, an ambition, and no mission. He was not yet popular enough to confront the Directory, yet he was too powerful to sit quietly in some Parisian mansion or suburb while waiting for an opportunity. The men of the Directory knew this, and they too looked for some way to rid themselves of the Corsican. It was here that Leibniz's vision was revived.

In the intervening decades, travelers to Egypt had written about their adventures. The Comte de Volney, born Constantin François de Chasseboeuf, but renamed in honor of the philosophe Voltaire, ventured to the Near East in the 1780s and wrote expansively of the untapped riches that Egypt offered. He concluded that "if Egypt were possessed by a nation friendly to culture . . . those monuments currently buried under the sand will be preserved as an invaluable resource for future generations." His book *Travels in Syria and Egypt* was widely read during the last days of the Ancien Régime. Flush with success, he went to Corsica to buy an estate, and there met a young Napoleon, whom he infused with a dream of conquest. Egypt, Volney hinted to Napoleon, was a chrysalis of greatness, if only there were a great man who could claim it.

Volney also influenced Charles Maurice de Talleyrand-Périgord, the pompadoured foreign minister already growing fat, who was then in the early stages of what would be one of the oddest, most influential diplomatic careers of the era. Talleyrand knew of Volney's ideas, because Volney was then involved in the politics of Paris. At the time, Talleyrand was jockeying for position in the shifting coalitions of the Directory. He looked at General Bonaparte as a tool and an adversary, and, like many, he wanted to remove Napoleon from Paris and send him to someplace far away. That place was Egypt.

In a series of meetings and memos between Bonaparte, Talleyrand, and the Directory in the spring of 1798, the rationale for an expedition was worked out. A strike against Egypt would be a blow against England, whose maritime expansion France dearly wished to halt. Egypt, said Talleyrand, was a land suffering from the tyranny of the Mamelukes, and what better way for the French Revolution to show that it was indeed ecumenical than to bring freedom to the oppressed? Egypt, said Napoleon, was a land of immense, untapped riches, and what better way to augment the power of France than to harness the wealth of the Nile? The Directory was convinced, and authorized Napoleon to raise a fleet at Toulon under conditions of secrecy and depart as soon as possible for Alexandria.[2]

Bonaparte was not then, nor would he ever be, a man of limited horizons. Others would have couched the Egyptian expedition in circumspect terms; for Napoleon, it became not just a conquest of a foreign land, but a mission ordained by God and demanded by civilization. On the eve of his departure in May 1798, he told Joséphine that he didn't know how long his mission would take—"a few months, or six years." Conquest would be followed by colonization, and colonization by further conquest—of the Holy Land, and perhaps even of India, where the English had recently evicted the French, and where the French maintained faint hopes of returning. Egypt would become an adjunct of French power, just as it had been the key to Roman power thousands of years before. Napoleon would become a new Caesar, or a new Alexander, and his landing at Alexandria would be the first step on that path to posterity.

But Alexandria in 1798 bore scant relation to the famous city of Alexander the Great. Inhabited by fewer than ten thousand people, it was a quiet port; save for a few desultory ruins on the outskirts, there were no signs of past glory. Napoleon landed his troops several miles from the city center. Facing little resistence, the army quickly made its way to Cleopatra's ancient capital and took control.

Napoleon had given his troops strict instructions. This was not an operation of pillage but, rather, a liberation. With only forty thousand troops, Napoleon planned to occupy a country with a population of more than three million people, hundreds of miles from his nearest supply lines. It would not take long for the expedition to consume the food and money it had brought from France, and if the occupation was to succeed, the assistance of the local population was imperative. Recognizing those constraints, Napoleon ordered his army to refrain from any theft, rape, or insult to the native population and their religion. Speaking on his flagship, *L'Orient,* Napoleon commanded his men not to contradict the basics of Islam. "Deal with them as you dealt with Jews and with Italians," he commanded. "Respect their *muftis* and their *imans,* as you respected rabbis and bishops. Show the same tolerance towards the ceremonies prescribed by the Koran that you showed towards convents and synagogues."[3]

Having occupied Alexandria, Napoleon reiterated this proclamation. He declared that he had come to liberate the people of Egypt from the tyranny of the Mamelukes and that he would be a friend and protector of Islam. He then left his fleet moored off the coast and advanced across the desert to Cairo. It was not the best time of year for an overland advance. Egypt in July is a cauldron, and the French troops wore wool uniforms and were unfamiliar with the terrain. The local Bedouin tribes harassed the soldiers and blocked up the wells, and it was days before the army reached the Nile. That they were able to make the journey at all in scorching heat with insufficient water is remarkable, but they marched on, and soon reached the outskirts of Cairo, at the northwestern suburb of Imbaba, where the Mameluke beys prepared to meet them.

On July 21–22, at the Battle of the Pyramids, Napoleon's army inflicted a crushing defeat on the mounted horsemen of the Mamelukes, a dynasty of soldiers and slaves that had once, centuries before, kept the Mongol invasion from streaming across North Africa, but were now split into rival factions. Napoleon exhorted his forces, "Go forth, and remember, that from the top of these monuments, forty centuries are watching us." Whether they were inspired by his words, they performed as he wished. The French lost several hundred men while killing several thousand. The remnants of the Mamelukes fled south, and Napoleon occupied one of the oldest cities in the world.

As he had in Alexandria, Bonaparte issued a proclamation to the people. He said that his only goal had been to liberate the city from the Mamelukes. He assured the citizens that they would not be harmed, that their homes would not be raided, that their women would not be violated, and that their religion would not be dishonored. Kind words notwithstanding, the people of Cairo did what people always do when faced with an alien army marching through their streets: they panicked.

Shaikh Abd al-Rahman al-Jabarti was an esteemed member of al-Azhar, the oldest university in the Muslim world. The mosque and school had been founded in the tenth century, and though al-Azhar had long since ceased being a seat of forward thinking, it was still a center of theology, and its shaikhs commanded respect. Jabarti chronicled the French occupation, and he described the terror that descended on Cairo in late July. People fled, and the French soldiers were not nearly as honorable as Napoleon had instructed them to be. "Night fell and the inhabitants of the city were in a great confusion and a fantastic uproar," Jabarti wrote. French chroniclers routinely portrayed Cairo in 1798 as decrepit, decaying, and declining. In reality, the city of 250,000 was thriving. Nominally under the aegis of the Ottoman sultan in Constantinople, Egypt was essentially independent, and Cairo was a thriving entrepôt of trade and industry. From the precincts of al-Azhar, three thousand students and teachers observed the French, and found them wanting in culture and manners.

Jabarti treated Napoleon's proclamation with contempt, in part for its content, but mostly for its terrible grammar. As if marking a slow schoolboy's paper, Jabarti went through the edict clause by clause to correct its many errors of spelling, syntax, and interpretation. He also lambasted the French for their lack of civility. When Napoleon's inner circle settled into the palace of Alfi Bey in the Ezbekkiyah Gardens, they "entered it, stepping on the carpets with their shoes and sandals as was their custom, since they never take off their shoes with which they tread upon filth, not even when they sleep! Among their repulsive habits also is their practice of spitting and blowing their noses upon the furnishings." Proud inheritor of more than a thousand years of learning, Jabarti saw the French as upstarts, and arrogant ones as well. They arrived trumpeting their civilizing mission, but from Jabarti's perspective, it was the French who needed civilizing.

This would not be the last time that Europeans and Egyptians stared at each other across a void of incomprehension. Both carried prejudices from centuries of contact and competition. Islam had emerged in the seventh century in the Arabian Peninsula as a new version of Judaism and Christianity, and though Islam taught tolerance for the other peoples of the book, the faiths coexisted uneasily. In the eleventh century, Western European nobles urged on by the Pope launched a crusade to retake the Holy Land from the Muslim dynasties that ruled Jerusalem, and for the next centuries, control seesawed. The Crusades ended in the thirteenth century with the eviction of the Europeans from the Holy Land, but they left a legacy of distrust and animosity that was rekindled when the Ottomans seized Constantinople in 1452. For the next two hundred years, the Ottoman Empire threatened the Venetians, then the Catholic monarchs of Spain in the sixteenth century, and finally the Austrians in the seventeenth century.

By the time Napoleon landed in Egypt, each culture was riddled with prejudice. The French, with the fervor of revolutionaries heralding a new age of man, saw the Egyptians and all "Orientals" as children in need of uplift; the grandees of Egypt and the Near East regarded Europe as a dangerous, barbarian adversary. When Napoleon occupied Cairo, he saw filthy alleys, crumbling buildings, and ruined monuments crying out for an injection of Western energy and innovation. When Jabarti looked at the French, he saw the latest wave of Western invaders who thought too highly of themselves, ignored the word of God, and were unable to recognize the marks of true culture.[4]

These impressions were strengthened by one of the odder aspects of the French expedition. In addition to 365 naval ships, hundreds of bottles of wine for Bonaparte's personal consumption, artillery pieces, blacksmiths, bakers, cooks, and thirty-six thousand troops for the expedition force, Napoleon brought 167 savants.

The mission of the savants was to survey Egyptian life, manners, agriculture, archeology, geography, cuisine, religion, literature, and whatever other desiderata of culture were deemed important. They were also to serve the needs of the expeditionary force, to work on such various tasks as purifying drinking water, finding local materials to manufacture gunpowder, and providing data and plans where needed. The average age of these "men of learning" was twenty-five, and many of them were students from the recently created École Polytechnique. Their specialties ranged from pharmacology to botany, architecture to musicology. They were bright and curious, but they were also very young and less learned than their sobriquet suggests. Still, among them were a few of the leading minds of France, including the mathematician and Jacobin politician Gaspard Monge, who at the age of fifty-two was a veritable elder statesman of the group.

Monge had met Napoleon in Italy several years before, and he accompanied the general on the flagship L'Orient. The two men spent many hours discussing the expedition and debating its goals. Monge could romanticize with the best of them. "Here I am," he wrote to a friend at the outset of the adventure, "transformed into an argonaut! This is another one of those miracles produced by our new Jason . . . who is going to carry the torch of enlightenment to a country which for such a long time has remained in darkness, and where he is going to spread republican philosophical thought whilst carrying our national glory even farther afield."[5] One of the mandates of the savants was to spread European knowledge to the Egyptians, as if they were missionaries of the Enlightenment and the Revolution. Once in Cairo, Monge and the 166 other members of the newly founded Institute of Egypt ensconced themselves in two Mameluke palaces and began to work.

They plunged into the exotic world around them, assembling instruments, hiring translators, and organizing survey parties. Monge initiated a study of the optics of desert mirages; others set up a printing press and published copies of a new journal. Even Jabarti was impressed. He was invited to visit the institute, and the members seemed to take great interest in his ideas. That openness and curiosity surprised him, given his initial opinion of the French. He admired the library they were assembling, and he was struck by the chemistry experiments that they performed for him, as well as the demonstrations of a new force that they called electricity.

Though the savants represented the kinder side of the invasion, Napoleon remained a conqueror, and when several factions in Cairo organized an uprising against his forces, he ordered his troops to shell the city from cannons mounted on the heights of the Citadel. The al-Azhar mosque was plundered, this after Napoleon had promised to respect the rights and religion of his new subjects. "They treated the books and Koranic volumes as trash, throwing them on the ground, stamping on them with their feet and shoes," Jabarti wrote. "They soiled the mosque, blowing their spit in it, pissing, and defecating in it. They guzzled wine and smashed their bottles in the central court." Far from friends, said Jabarti, they behaved like "the host of Satan."[6]

Jabarti may have exaggerated. He was, after all, a partisan for his country. The French, however, were probably no more brutal than previous regimes. A rebellion had occurred, and Napoleon suppressed it with the same ruthless efficiency he demonstrated throughout his career. Even so, he was left in an untenable situation. His fleet had been destroyed near Alexandria by the English Admiral Horatio Nelson, and his remaining forces were trapped in a hostile country, with few supplies and a dwindling source of money. He needed to secure Cairo and devise a new strategy. Retreat was now impossible, and the only option was to advance into Syria and the Holy Land.

Napoleon had made no provision for this type of incursion. Over the centuries, countless invaders had crossed the desert between Cairo and the Sinai Peninsula and then into the Holy Land. But the French did not know the best routes, and in late 1798, Napoleon went on a scouting mission. A larger force first made sure that there were no hostile troops lingering in the region, and then Napoleon set out, accompanied by several of the savants, a small company of troops, engineers, and a group of surveyors led by Jacques-Marie Le Père. They headed east. Napoleon wanted to get a feel for the terrain before he ordered a full-scale deployment. He also had a special project for Le Père.

Of the many instructions given to Napoleon by Talleyrand and the

Directory, there was one that seemed improbable at the time. "The general in chief of the Army of the Orient will seize Egypt; he will chase the English from all their possessions in the Orient; and he will destroy all of their settlements on the Red Sea. He will then cut the Isthmus of Suez and take all necessary measures in order to assure the free and exclusive possession of the Red Sea for the French Republic." Listed along with several other instructions, this directive made the task seem almost pedestrian—"cut the Isthmus of Suez." The Directory had wanted Napoleon out of Paris, and, much like the proverbial Labors of Hercules, joining the Red Sea to the Mediterranean should have kept Bonaparte occupied until the members of the Directory had gone to their graves.

To honor the letter of these instructions, Napoleon turned to Le Père. Various canals had existed in antiquity. As Le Père knew, these had been described by ancient geographers, and traces still remained in the desert. But most of those earlier canals were indirect routes. Rather than one immense trench linking the Red Sea directly to the Mediterranean, the old waterways took a circuitous path from the port of Suez to an area near the Bitter Lakes, and from there, west across the desert to the Nile. The Nile was then used for transport to and from the Mediterranean, though ancillary canals were needed to allow navigation through the delta.

At Napoleon's urging, Le Père began with a blank slate. Rather than trying to copy the ancient routes, he assessed every possible permutation. Starting in December 1798, he made four separate trips. He surveyed the hundred-mile-long stretch of isthmus between the two seas, and he charted the terrain between Suez and the Nile. He judged the port of Suez a depressing place, especially in light of its noble history. The northern part of the isthmus was a wasteland of desert and marsh surrounding the long-abandoned branch of the Nile at Pelusium. In addition, the region was without fresh water, and in those weeks of December and January, it was freezing cold at night, and dangerous as well. Bedouin tribes had carved up the various regions among themselves, and the French were constantly on alert for attacks.

That made a difficult task almost impossible. The tools for topographical surveys in those years were crude, and often required more legwork than scientific knowledge. Teams of men had to place stakes throughout the survey area, and then compile measurements. That meant time and meticulous deployment. But, with the constant danger of attack and capricious guides, the work was rushed. Hurried and harassed, Le Père made a critical mistake and concluded that the waters of the Red Sea at high tide were more than thirty feet higher than the waters of the Mediterranean.

Le Père's measurements did not pose a problem for a canal from Suez to the Bitter Lakes. For that southern portion, he reported, there would be no risk that the land would be inundated, because the levels were essentially the same. Except for several topographical obstacles, such a canal would be feasible. But in the north, Le Père decided, there would be an insurmountable problem. Because of the difference in level, the waters of the Red Sea would flood the marshy flatlands to the south and east of Lake Manzala, inundating the area with a thin layer of water and making it difficult to navigate. Even if a route could be created through the flooded region, there was no place for a port. He concluded that a direct canal was impossible, and his findings were accepted—not without some question, but in the end his survey became the official conclusion of the Institute of Egypt. Under the title "Report on the Communication of the Indian Ocean with Mediterranean by Way of the Red Sea and the Isthmus of Suez," his results were incorporated into the magisterial *Description de L'Égypte* that the institute began publishing in 1809.

Of course, by that time, Napoleon cared little for the fate of Egypt. His fleet destroyed by a superior English commander, his advance into Syria stymied by a dearth of troops and overextended supply lines, Napoleon slipped out of the country in August 1799, leaving his army and his generals to face an increasingly restive populace and an English invasion. As for the canal, he brushed off Le Père, not harshly but with flattery. "It is a great thing you propose," he told the chief engineer. "Publish a memoir and force the government of Turkey to see that its glory and its interest will be served by its completion." Having treated Egypt as a path to greatness, he recommended that Le Père do the same.[7]

Napoleon had mused to Joséphine that he might be gone six months or six years, and in the end it was closer to the former than to the latter. He returned to Paris at the end of 1799, and within weeks mounted a coup that brought him to power. Somehow, he managed to portray his year in Egypt as a glorious success, and he pointed to the achievement of the savants and the institute as proof that he had brought light to the darkness of the Orient.

Though the French were evicted from Egypt in 1801, the work of the institute continued in France. That was only one of the legacies of Napoleon's expedition. The vacuum left by the defeat of the Mamelukes was filled by Muhammad Ali, an Albanian mercenary in the service of the Ottoman sultan. Though it would take Muhammad Ali several years to consolidate his hold, he became known as the founder of modern Egypt. Without Napoleon, however, there would have been no Muhammad Ali; and without Muhammad Ali, it might have been many decades before Egypt began the arduous, disorienting, and not entirely successful process of modeling itself along the lines of a European state. Without Napoleon, there would also have been no Institute of Egypt, and no massive nine-volume publication on all aspects of Egyptian culture. And that would have meant no French fascination with Egypt, and no love affair with the land of the pharaohs.

Given that Bonaparte went on to conquer most of Europe, that he was crowned emperor by the Pope himself, that he overturned much of the old order, and that, had it not been for his overreaching in Russia, he might have succeeded in establishing a French imperium stretching from the Atlantic to the Caucasus and from the North Sea to the Mediterranean, the Egyptian campaign can seem relatively insignificant in comparison. But even if it was only a prelude to Napoleon's career, its effects on the eastern Mediterranean, on Egypt, and on the future were profound.

The expedition instilled in France a passion for all things Egyptian. The well-to-do of Paris and the provinces adorned their homes with vases, candelabra, and linen decorated with ancient Egyptian motifs. Obelisks, sculpture, and public monuments copied Cleopatra's Needle and the temples of Upper Egypt at Philae, Edfu, and Dendera. Centuries of Christian and Muslim Egypt were of no interest to the French public, but ancient Egyptomania thrived. The lust to uncover the ruins of Egypt reached a fever pitch by the 1820s and captured the imagination not just of the French but also of the British and the Germans. Archeology, which had existed only as a rudimentary science, blossomed, and by the 1820s, when the politics of Egypt had stabilized, European men of learning flooded the country, each trying to claim a piece of glory by unearthing some tomb or temple.

One of the pivotal discoveries of the savants was a large stone found in the coastal city of Rosetta. Inscribed in hieroglyphics and in two forms of Greek, the stone provided crucial clues to European linguists who had been trying unsuccessfully to read the inscriptions of the ancient Egyptians. Even with the stone, however, hieroglyphs did not yield their secrets so easily. In France, the race to crack the hiero-glyphic code took on national importance as the precocious linguist Champollion competed against an equally brilliant and ambitious En-glishman named Thomas Young. To France's pride, Champollion made the first breakthrough when he figured out that hieroglyphics were nei-ther purely phonetic nor purely symbolic. That discovery, in 1822, was as dramatic in its way as an archeologist stumbling upon a buried trea-sure. Suddenly the culture of ancient Egypt was open to inspection, and the flow of new knowledge and new information cemented the passion for things Egyptian.[8]

With the tumult of the Revolution and the Napoleonic Era now passed, France in the 1820s was stuck in a political limbo. There was a king, but no universally accepted ideology of royalty. There were repub-licans, but no agreement on democracy. And most of all, there was a culture of progress on the one hand, and romanticism on the other. These forces pulled in opposite directions. Romanticism looked to the natural world, and sadly concluded that beauty and glory were fading as mankind advanced. The cult of progress celebrated the industry of the modern world, and declared that the best was still to come. Egypt offered the chance for reconciliation. The wonders of Egypt, in the eyes of Western Europeans, lay buried by centuries of decay. The country had been great, but it was no longer. Yet Egypt also offered a new horizon. One could idealize its past, and also believe that it could once again shine. It was a land where the cult of progress and the cult of romance could walk hand in hand.

Napoleon's rapid rise and sudden fall added to the lure of Egypt. Somehow, the notion that France's destiny was linked to the Nile took hold in the French imagination. That the expansion of the British Empire also seemed to hinge on control of the eastern Mediterranean only intensified the French desire to be the pre-eminent power there. The first great civilization had emerged in Egypt; the Greeks had bor-rowed from the pharaohs as prelude to their own efflorescence; the Roman conquest of Egypt directly preceded the rise of Augustus; and then Napoleon went there to achieve his destiny. Clearly, or so the French thought, Egypt was a source of grandeur.

But what was to be done about Egypt as it actually was in the first decades of the nineteenth century? That was answered by the Egyp-

tians themselves, or at least by their ruler, Muhammad Ali. He saw
Egypt much as the French did: its glories were past, and perhaps they
also lay ahead, but if that was to happen, Egypt would have to change.
And who better, thought the transplanted Albanian soldier, to assist in
that change than the French?

Napoleon was, therefore, an inadvertent matchmaker. He began an
affair between France and Egypt that placed Egypt once again at the
center of the world. Before that could happen, however, the courtship
needed to become more intimate, and it did, in 1833, when a group of
missionaries landed at Alexandria proclaiming the union of East and
West.

# CHAPTER THREE

# INDUSTRY AND THE SAINT-SIMONIANS

THE FRENCH REVOLUTION unleashed a torrent of energy in France. Napoleon Bonaparte was one result; messianic fervor was another.

A revolution that declares the old order corrupt and says that the world can be made new rests on a good degree of optimism. It demands a belief that a better life is possible, if only the right formula can be found. Having proclaimed the dawn of a new age, the French Revolution fertilized the imagination. Suddenly, it seemed, everyone had an idea for how to improve life. And one of those ideas came from an aristocrat named Claude-Henri de Saint-Simon.

Scion of the house of the Duc de Saint-Simon, who had written a salacious memoir of the court of Louis XIV at Versailles, the young Henri briefly fought on the side of the Americans in the War of Independence. Reflecting on that experience later in life, he wrote, "I realized that the revolution in America signaled the beginning of a new political era, that this revolution would necessarily bring about major progress in general civilization, and that in a short time it would cause great changes in the social order which then existed in Europe." That was, in fact, what happened, and when the French Revolution began, the twenty-nine-year-old Saint-Simon eagerly joined the ranks.

But he fell out of favor during the Terror, largely because of ill-timed profiteering. One of his more far-fetched schemes involved a plan hatched with Talleyrand to buy the Cathedral of Notre-Dame and then strip off its roof, melt it down, and sell the metal. He was arrested, but spared execution, and though he kept his head safe from the guillotine and outlived Robespierre, when he was finally freed he had lost most of his credibility as a revolutionary. He prospered nonetheless, dabbling in real-estate deals, until he had a philosophical epiphany and began to write.

In the years before his emergence as a thinker, Saint-Simon became involved with a new school in Paris, which was to be one of the most consequential acts of his very consequential life. Created in 1794, the École Polytechnique was a manifestation of the revolutionary belief in the perfectibility of mankind. The school was established to train young men, usually teens, in the arts of sciences and industry. A harbinger of the state funding of education that became a hallmark of France and the rest of Europe in the nineteenth century, the Polytechnique provided grants to students who showed scientific promise. Within a few decades, the school exceeded its founders' ambitions and was widely heralded as a center for innovation.

By then, Saint-Simon had transformed himself. When he was forty-two, he published his first work of philosophy. He called for the establishment of a religion of science with Sir Isaac Newton as its patron saint. Over the next two decades, he published numerous treatises and gathered around him a group of energetic followers dedicated to the propagation and implementation of his ideas. They began to spread the gospel of Saint-Simon, one of whose tenets was that the world should be linked by canals.

The passion for canals was part of an overall philosophy of industrialization. Developed in fits and starts over the first decades of the nineteenth century, Saint-Simon's system called for a society led by scientists and industrialists. Whereas most of Western history had placed the greatest value on those who own property while others farmed it, Saint-Simon spoke for a new generation of Europeans who prized work. He held that the best life was one devoted to industry, to building things and making things. Philosopher-industrialists should lead society, and the bottom of the social pyramid should be occupied by idlers, by people who lived off others or had no vocation. Only through hard work and industry, said Saint-Simon, could there be progress, and progress was man's destiny.

These beliefs were radical. They challenged the doctrines of the church and the old regime. Most traditional church theology hadn't been based on the notion of progress, and by the nineteenth century, the French ecclesiastical establishment had fallen behind secular thinkers. Saint-Simon built on several strains of Enlightenment thought, and crafted them into a doctrine that the future will always be better than the present. He then applied these ideas to the nascent industrial order emerging in postrevolutionary France. He rejected the notion that humans live and die according to God's plan and that the ultimate goal is heavenly rewards. Instead, he enshrined earthly progress as the ultimate good. He recognized, however, that progress demands leaders. Certain types of people are more adept at leading the onward march—namely, scientists and engineers. Saint-Simon stressed that it was society's job to mold such people at places like the Polytechnic School. There, future leaders would learn how to manipulate nature and shape the earth to suit the needs of mankind. Once society was led by these types, there would be no limit to what could be achieved, and no barriers to limitless prosperity, universal peace, and international harmony.

Saint-Simon was not a rigorous, organized thinker. He flitted from idea to idea, backtracked on some, and reworked others. He enjoyed whiling away hours in Left Bank cafés, talking with disciples and students, and his books often read as disjointed compilations of compelling ideas. Soon enough, however, a rough consistency emerged. His philosophy became known as "positivism," which among other things suggested that history was headed in a positive direction and that science and industry would lead the way. The most celebrated disciple of Saint-Simon was the young Auguste Comte, who broke with his master and went on to develop positivism as a more coherent philosophical system whose influence on nineteenth-century Europe and the United States is difficult to overstate.

Saint-Simon, however, shifted gears later in life. He became more mystical, and more religious. He tried, with typical bravado, "to systematize the philosophy of God." He asserted that civilization consists of dichotomies, especially between the temporal and the spiritual. The needs of the spirit, he went on, are distinct from the needs of the flesh. Historically, all civilizations had been bedeviled by that split. In the Catholic Church, the flesh was seen as sinful and the spirit as potentially noble, and that duality continued to plague Western civilization. To close the gap, Saint-Simon proposed to do nothing less than create a new religion, which he called, simply enough, the New Christianity.[1]

Concerned that positivism alone might perpetuate the duality that had crippled human history, Saint-Simon gave God, rather than the abstract notion of progress, the seat at the head of the table. He began to speak more about love and less about science. He evoked the language of St. Paul, and preached the golden rule to treat others as you would have them treat you. But, however much this theology stemmed from Christianity, it was a new form of it, and, unlike the patriarchs and the Vatican, Saint-Simon stressed that this world, not the next, was the only relevant arena for human destiny.

His disciples kept his ideas alive, and as disciples do, they fought over his legacy and then invented variations. After Saint-Simon died in 1825, his leading students quarreled over his mantle, and several factions emerged. The most dynamic, and most peculiar, of these was led by Barthélemy-Prosper Enfantin. Combining the rites of pagan cults with the rationalism of the Enlightenment, Enfantin initiated a chain of events that led directly to the creation of the Suez Canal.

A contemporary of Comte, born in 1796, Enfantin was educated at the Polytechnic School and briefly served in the last armies of Napoleon. An unsatisfying career as a roving wine-merchant led him back to Paris, and by the early 1820s he had made his way into Saint-Simon's inner circle. As befits his name, Enfantin was a lithe, graceful, effeminate man, who later developed a predilection for dressing in woman's clothing. More than anyone, he embraced the New Christianity. During a series of symposiums among the Saint-Simonians in 1829, Enfantin explained his views to the gathering: "We come to proclaim that mankind has a religious future; that the religion of the future will be greater and more powerful than all those in the past; that it will, like those which preceded it, be the synthesis of all conceptions of mankind, and moreover, of all modes of being. Not only will it dominate the political order, but the political order will be totally a religious institution; for nothing will be conceived outside of God or will develop outside of His law. Let us add finally that this religion will embrace the entire world because the law of God is universal."[2]

Saint-Simon had announced that society ought to be reorganized around a simple axiom: social institutions should be dedicated to the task of the material and moral improvement of the largest number of people. That was a bold idea, but it was Enfantin who added the additional twist that has since become an adage: "To each according to his capacities, and to each capacity according to his works." That meant a society based on individual merit, in which those of greater capacities would naturally lead those of less. The greater good would always be the North Star, but some would have more sway than others.[3] In essence, Enfantin blended Plato's ideal of the philosopher king with what would later be known as the superman pathology of Nietzsche. It was a potent brew.

Enfantin was not modest. He described himself as "one of those lovable beings who is followed." He gained the loyalty of the bulk of the Saint-Simonians, and he quickly established himself as a creative thinker in his own right. He took up where his teacher had left off. Saint-Simon had declared, "The Golden Age is not behind us; it is ahead; it is in the perfection of the social order; our fathers never saw it; our children will arrive there one day; and it is us who will pave the way." But, unlike his master, Enfantin believed that the paving stones would have to consist of a good deal of metaphysical mystery. Starting with the notion that all civilizations are crippled by dualism, Enfantin wove a theology that covered all aspects of human existence. Not only was there a fissure between the temporal and the spiritual, between the world of Caesar and that of God, there was an even wider chasm between the masculine and feminine, men and women.

That split manifested itself in the inequality of social relations, which in turn impeded the ability of societies to grow as rapidly as progress demanded. Enfantin asserted that the promise of the New Christianity could only be realized if sexual relations were reformed. After the 1829 Parisian symposiums, Enfantin formalized his break with the rest of the Saint-Simonians and moved his followers to the suburb of Ménilmontant to form a new church. There, he became known as "Père" Enfantin, the "Father." His followers were called "apostles." Satellite churches were set up throughout the country, in places such as Toulouse and Lyon. Enfantin called for the abolition of prostitution, and he insisted that women should have the right of divorce, that they should enjoy the same legal guarantees as men, and that husbands should not have the right to the property of their wives. "Our apostolic work," Enfantin wrote, "consists principally in the Appeal to Women and in the Rehabilitation of the Flesh through the political organization of industry and the creation of a new cult." In addition to his sexual radicalism, he also seems to have considered himself the reincarnation of Jesus Christ, or at least "the instrument which God had sent to change the life of all men and women, just as He had sent Jesus, Mohammed, Moses and others to change the world in other epochs." And in case this was not a sufficiently lofty peer group, Enfantin also considered himself a descendant of St. Paul.[4]

At Ménilmontant, this latter-day saint set himself up as the leader of a new order. Life in the community was regimented. Like monks, disciples were required to be celibate. Their morning began at 5:00 a.m., and for the rest of the day, prayer and lectures were interspersed with physical activities that ranged from cooking to masonry. Gardening received special attention. Every aspect of life was designed to reflect the overall philosophy. Songs were composed to accompany activities, and the lyrics praised Enfantin and celebrated God and the new world that was being created. For residents used to a life of privilege, this pseudo-monasticism was a break from society; for many, it may have been the first time they had washed their own clothes. Of course, none of them had worn clothes quite like those Enfantin required them to wear.

In an acute example of how clothes make the man, the Father ordered his apostles to wear vests that buttoned at the back, flared trousers that could easily be mistaken for a twentieth-century woman's skirt, a brightly colored sash tied at the waist, and stockings that extended above the knee. White scarves were to be tied around their necks and under their beards, and, for travel, the disciples wore red berets that covered their shoulder-length hair. Some of it was aesthetic, but some of the custom-designed apparel had metaphoric significance. To secure the vests at the back required assistance, to symbolize, said Enfantin, the dependency of each individual on the kindness of others.

But Enfantin was not content simply to retreat from the bustle of Paris and redesign the couture of his followers. He meant to be a force in society, and that demanded engagement with it. He wanted to change the world, and he couldn't do so from a private estate in a suburb in northern France.

Eighteen thirty-two was a difficult year. Enfantin and his leading disciples were put on trial for offending public morality. Their stance toward the liberation of women, and the rumors that, despite claims of celibacy, they engaged in orgies, agitated their less flexible neighbors. Having trumpeted Enfantin's philosophy in various publications, the community had become something of a tourist attraction, and that both increased Enfantin's visibility and incensed the guardians of law and order. He was briefly imprisoned, and the experience unnerved him so much that he looked beyond France for the next phase of his master plan. Having explored the dichotomy between spirit and flesh, and between male and female, he decided that history demanded the union of the two, and the venue he selected for that marriage was Egypt.

With an eye toward the dramatic gesture, Enfantin declared 1833 the Year of the Mother, and then organized an expedition that mimicked Napoleon's in ambition, though not in size. "I heard from my prison," he wrote to a friend in January 1833, "the Orient waking up and crying out. . . . The Nile has broken through its dikes, and it gushes out farther than it ever has before, carrying the seed that Napoleon scattered on its banks, which Muhammad Ali now fertilizes. . . . The great communion draws near; the Mediterranean will be beautiful this year."[5] By early spring, Enfantin had assembled a group of several dozen men and a few women. Most of them were engineers trained like Enfantin at the Polytechnic. He also sent a delegation of his community to Constantinople. Calling themselves "The Companions of the Woman," they offended the Ottoman authorities by aggressively touting their views about sexual relations and preaching the gospel of the New Christianity. Unwisely, they declared that a new day, presumably one that would dramatically alter the role of the sultan, was nigh. Narrowly avoiding prison, they were told to leave Turkey, and they met up with Enfantin and his party in Alexandria in the fall.

The Father's motives were a combination of the sublime and the mundane. He said that Egypt was the "nuptial bed" that would allow for the symbolic union of East and West. The East represented the female principle, the West the male. The East, for Enfantin and his disciples, was that portion of world civilization that emphasized the universal Mother, the earth goddess. The West signified the Father and the spirit of intellect, reason, and creativity. On its own, each was lacking a key ingredient, and for that reason, the world was locked in struggle. Only when these two were no longer divided by centuries of human pettiness and distrust would universal harmony and prosperity be more than a utopian dream. By making the journey to the East, Enfantin planned to heal the fissure and thereby inaugurate a new age of progress, industry, and peace. Once the marriage was consum-

mated, the Year of the Mother, dedicated to women, would be succeeded by the Year of the Father, dedicated to industry.

To clarify the goal, Enfantin or one of his followers composed a brief
poem:

> *It is we who will make*
> *Between ancient Egypt and old Judea*
> *One of the two new routes from Europe*
> *To India and China*
> *Later, we will pierce the other*
> *At Panama*
>
> *Suez*
> *Is the center of our life's work*
> *There, we will do something*
> *That the world will witness*
> *In order to confess that we are*
> *Males*[6]

A canal through the Isthmus of Suez would consummate the marriage, because then the waters of the Mediterranean would mingle
with those of the Red Sea. The Mediterranean was the sea of the West,
of the male Occident. The Red Sea was of the female Orient. Once the
canal was cut, the two seas, representing the two eternal forces of life,
would join, and the duality that had sundered the planet would be no
more.

It is hard, at two centuries' remove, to take Enfantin altogether seriously. It is easier to dismiss him as a blowhard than to consider him a
visionary. His language wasn't simply florid; it was imbued with a sense
of destiny and metaphysical purpose that seems more suited to mystics
or madmen. But although he was more emphatic than many of his contemporaries, he was not completely outside the mainstream, at least
not in his ambitions or his belief that he had been marked by history to
perform monumental deeds. Europe in the first part of the nineteenth
century was full of people who spoke in exaggerated terms about their
own fate and the history of the world. Indeed, Enfantin's way of
expressing himself was not so different from that of the late-eighteenth-
century founders of the United States or the revolutionaries of France.
His specific ideas about women, and about the East and the West, may

have been eccentric, and the odd dress code struck his contemporaries as, well, odd, but the sense of moment was not especially remarkable in a world occupied by Romantics such as Byron and Chateaubriand, by Jefferson and Napoleon Bonaparte. Later, when Lesseps began his mission, he spoke in similar, though less metaphysical, terms.

His reputation preceding him, Enfantin and his coterie disembarked in Alexandria to a crowd of onlookers, who must have been somewhat bemused by the peculiar appearance of the travelers. After the hostile reception in Turkey, Enfantin and his acolytes had toned down their appearance. Beards were trimmed or shaved off, and the outfits were altered or muted, but they still made a spectacle. Alexandria was a far different place from the decaying port occupied by Napoleon three decades before. Under the rule of Muhammad Ali, Egypt was gradually being integrated into the world commercial system. Alexandria had linked Asia and the Mediterranean for most of antiquity, and after centuries of neglect and decline, it was once again thriving. The city that greeted Enfantin had seen a recent influx of French merchants, English entrepreneurs, Austrian middlemen, Greek traders, and a crazy quilt of other Europeans, and they conducted themselves with the jaded sophistication typical of multicultural cities devoted to commerce.[7]

By the time he arrived in Egypt, Enfantin had undergone a subtle shift. He continued to speak in terms of the nuptials between East and West; he still saw himself as the vehicle of the virile Occident who would impregnate the Orient; and he talked of his mission to Egypt as "a new crusade, guided by the eternal Star transfigured in the Epiphany."[8] But he also focused more on doing and less on philosophizing. The pragmatic component of his "religion" needed tending, and his apostles were trained engineers. Philosophy aside, Enfantin traveled to Egypt because he had a plan: to convince Muhammad Ali to use the expertise of the Saint-Simonians to modernize the country.

Muhammad Ali closely resembled the Western image of an Eastern potentate. Bearded, somewhat short, usually adorned in billowing yet simple robes, he greeted petitioners while reclining on a divan. Slaves attended to him and ministers cowered in his presence. He was surprisingly soft-spoken and invariably polite, with a gentle, melodious voice. Visitors were often struck by his eyes, which seemed at once playful and ominous. Like many self-made men, he possessed an iron resolve, and his adversaries frequently blinked before he did. In 1811, he invited the remaining members of the Mamelukes to a sumptuous banquet in the Jewel Palace at the Citadel, and as they were relaxing in a gustatory haze, his guards sealed the room and massacred all but one, who jumped into an alley through a large open window and escaped on his horse. Having killed the remnants of the old ruling class, Muhammad Ali then appropriated their lands.

While the pasha had since mellowed, he was still the absolute master of his realm. Anything of consequence required his approval. Enfantin could only realize his goals if Muhammad Ali permitted it. At least there was some precedent. The pasha had brought a number of Europeans into his service, employing one as director of irrigation and another as head of health services. He recognized that Europe had surpassed the Near East in its ability to grow food, build cities, produce revenue, employ soldiers, and arm them with advanced weapons. By the time Enfantin arrived, Muhammad Ali had consolidated his hold on Egypt and had mounted a serious challenge to his nominal overlord, the Ottoman sultan. He sought European assistance to further his ambitions. The English supported the sultan and were wary of aiding Muhammad Ali; the French, out of their love for Egypt and their rivalry with England, rushed in to provide whatever help the pasha desired.[9]

But Enfantin didn't just want to enter Muhammad Ali's service. He also wanted to obtain contracts known as "concessions," which gave the concessionaire a limited monopoly. A concession gave a group or an individual the right to undertake a specific project, whether it was a dam across the Nile or a railroad from Alexandria to Cairo. The concessionaires would then form a company to implement the project, and they would capitalize it themselves. They would also be entitled to most of the profits, with a stipulated amount reserved for the Egyptian government. The system of concessions benefited both parties. Muhammad Ali obtained public works for a minimal cost, and European entrepreneurs were able to set up businesses. Such, at least, was the theory. Unfortunately for Egypt, as for countless other non-European societies in the nineteenth century, these concessions often came with hidden costs for the ruler granting them.

Enfantin passed a few pleasant weeks in Alexandria, mingling with the expatriate community, before heading to Cairo, where he was warmly entertained by the French consul, Monsieur Mimaut. Though politeness demanded Mimaut's hospitality, it was not an unpleasant duty. Enfantin was a colorful personality with a large following among powerful French industrialists and businessmen. He discussed his ambitions with Mimaut and the consular staff, and his sermons charmed dinner-party guests. Never one to keep his ideas to himself, he told any who asked about his plans for railroads, dams, and canals. He also talked about Egypt as the perfect setting for the next phase of world history. In Europe, he said, political power was locked in a dele-terious struggle with industrial power. He pointed, as many others did, to the tumultuous events of Paris in 1830, when King Charles X was removed from his throne in the "bourgeois revolution." In Egypt, Enfantin believed, there was little tension between politics and indus-try, and with a man such as Muhammad Ali in control, industry could coexist with political power.

Enfantin was soon making the rounds in Cairo. He needed to arrange an audience with Muhammad Ali, and that took time and con-nections, and perhaps even a few selective bribes. He met chamber-lains and consuls, and he lobbied the Europeans in the service of the pasha. One of these was Maurice-Adolphe Linant de Bellefonds, who had been in Egypt for more than a decade and had ascended to the position of chief engineer. The projects being proposed by Enfantin required Linant's support, or at least his lack of opposition. Linant's area of expertise was irrigation, and he was a keen student of canals. In early January, on Linant's urging, Enfantin took a brief trip to the famed Suez. Accompanied by his disciples, he went first to Damietta, then to the Bitter Lakes, in the middle of the isthmus, and then south, to the port of Suez, before returning to Cairo.[10]

There was one other Frenchman whom Enfantin encountered in these months, one who would in time become his most avid convert— not to the New Christianity, but to a plan to cut a direct canal linking the Mediterranean and the Red Sea. In later years, however, Ferdinand de Lesseps did everything in his considerable power to distance him-self from Enfantin, and he successfully fought to claim the legacy of the Suez Canal for himself and himself alone. During his stay in Egypt, Enfantin made no secret of his desire to build a canal through the isth-mus. Had he known that one of his listeners, a polite young vice-consul, would one day appropriate his idea and achieve international fame and fortune, he might have kept his own counsel.

In early January 1834, the pasha agreed to receive Enfantin. Muhammad Ali, by all accounts, listened attentively to the proposals, though he had been briefed on them in advance. He endorsed the idea of irrigation works along the Nile because he was eager to construct a massive barrage north of Cairo that would allow for water to be stored and released as need be. The Nile was notoriously fickle in its flooding, and for millennia, Egyptian rulers had tried unsuccessfully to moderate the floods so that they would be able to predict annual crop yields. Muhammad Ali needed cash crops to raise money to finance his modernization schemes, but unless the floods could be regulated, there was no way to rationalize long-term planning. For that reason as well, he wanted to link Alexandria more tightly to Cairo. Rather than collect a tax quota from the merchants and leave them to their business, he planned to transform Alexandria into a national trade gateway. From its port, the cash crops of the Nile would be exported to Europe, and imported weapons, machines, and materials would arrive at Alexandria before making their way south through Cairo and into Upper Egypt. Each stage of this loop would be managed by individuals directly accountable to Muhammad Ali; he would dictate prices, taxes, and salaries in order to generate maximum revenues for his treasury.

But the pasha was not enthusiastic about a Suez Canal. As far as he could tell, no good would come of it, not for Egypt and not for his dynasty. A canal to the east of the Alexandria-Cairo corridor built by Europeans would simply allow the powers of Europe to trade with the Orient without having to pass through areas controlled by Muhammad Ali. He saw a day when he and his heirs would be cut out entirely, or, worse, undercut completely. As long as trade had to make its way from Alexandria to Cairo and then overland from Cairo to the port of Suez, he could control it. His customs officials could charge duties and take a percentage, and Egyptian middlemen could become rich from the proceeds. That is how he preferred to keep it, and that is what he conveyed to a disappointed Enfantin.

Undeterred, Enfantin took what was offered. He went to work on organizing the Nile barrage, and with Linant he developed preliminary sketches for improving the Mahmoudiah Canal, which connected Alexandria to the Nile. At the same time, he refined plans for Suez. One no from Muhammad Ali was not enough to stop him from pursuing his dream. Work on the barrage proceeded quickly, aided by the tens of thousands of corvée laborers supplied by Muhammad Ali. Enfantin also established schools modeled on the Polytechnic in Paris. Muhammad Ali was a strong advocate of improved education for Egyptians, especially technical education that would give his countrymen the knowledge to build bridges, waterworks, roads, and weapons. Though Enfantin was no fan of weaponry, he happily placed Saint-Simonians in positions of authority. Soon, his apostles were founding schools of their own. Embraced by the pasha, they slowly pulled away from Enfantin. Though most continued to pledge their allegiance to him, several decided to stay permanently in Egypt.

Except for the canal, the mission to Egypt was a success. True, the marriage that Enfantin had talked so much about never quite came to pass. But in other respects, Egypt fulfilled his expectations and satisfied the ambitions of his followers. To celebrate their good fortune, the Saint-Simonians in Egypt organized a party on August 15, 1834. Enfantin chose that date because it was Napoleon's birthday. Seeing himself as a reincarnation of Christ, Enfantin claimed the mantle of Napoleon as well. The festivities were meant to link the Saint-Simonians in Egypt to the legacy of Bonaparte, and just as Napoleon had organized fireworks displays over the skies outside of Cairo, the Saint-Simonians staged a fête that drew the attention it was designed to attract. The workers on the barrage were given the day off, and an official delegation from the court of Muhammad Ali, along with the representatives of France and other governments, toured the works and the encampment. In a specially outfitted tent, Enfantin hosted a lavish lunch with three large tables stacked with delicacies. The nineteen guests were treated to sixteen bottles of champagne, fifteen bottles of Burgundy, and a dozen from Provence. The luncheon ended with the ceremonial laying of the cornerstone for a new school of engineering.[11]

That day was the high point of Enfantin's influence in Egypt. Soon, the barrage encountered the inevitable delays and problems that beset most major engineering projects, and the loose coalition of Saint-Simonians, Egyptian officials, and French in the service of the pasha began to disintegrate. The Father had neither the patience nor the temperament to deal with these issues, and he decided to return to France in 1836. He never returned. But he was not done with Egypt, nor had he given up on the idea for a canal.

But when Enfantin finally turned his attention to the canal once again, he was confronted with a man whom he had known in Egypt only as a diligent representative of the French government. Ferdinand de Lesseps was the official delegate of France for lunch that day in Enfantin's tent. Years later, out of work and in disgrace, Lesseps remembered the dreams of Enfantin, and he made them his own.

CHAPTER FOUR

# A Man, a Plan, a Canal

*Some men are born great,*
*Some achieve greatness,*
*Others have greatness thrust upon them.*

—William Shakespeare, *Twelfth Night*

I N December 1804, the French consul in Cairo went home to visit
his wife and brief the emperor. Though Bonaparte had slunk out of
Egypt, he emerged with his reputation intact, and after seizing
power, he rewarded the minister Talleyrand with rank and position.
Egypt was no longer a high priority for France, but the strategic contest
that had led Napoleon there in the first place remained unresolved.
The British Empire still craved primacy in Egypt, and Napoleon and
Talleyrand sought to prevent that. They could not spare troops, but
they did delegate a seasoned diplomat named Mathieu de Lesseps to
do what he could to thwart the British.

As consul, Mathieu performed his task with élan. He befriended
Muhammad Ali and helped the charismatic, ruthless leader gain control of the country. Though the situation in Cairo was fluid in the fall
of 1804, the pasha had defeated most of his immediate adversaries.
One reason for his success was Mathieu de Lesseps, who identified
Muhammad Ali as a man capable of dominating the warring factions
in Egypt. The alliance between France and Muhammad Ali was
cemented, as most things are, by a mundane event. At a banquet
thrown by the French consulate, some silverware disappeared.
Muhammad Ali fell under suspicion. Stealing French cutlery was not
a cardinal offense, but it wouldn't help the Albanian's cause. Mathieu
came to the pasha's defense and declared him innocent. A seemingly

38

minor gesture, it preserved Muhammad Ali's honor and reputation, and that was not a minor matter.[1]

The story may be apocryphal, but the solid relationship between France and the new ruler of Egypt was anything but. Mathieu had earned some rest, and he seems to have spent the time well. Nine months later, he was back in Egypt, and his wife, Catherine Grivégnée y Gallegos, gave birth to a son, whom she named Ferdinand.

It may be, as Shakespeare wrote, that some men are born great, but for Ferdinand de Lesseps, it was much more a case of achieving greatness by dint of will and force of circumstance. To be sure, he was born in elevated circumstances. His father, though absent for long stretches of time, was a respected diplomat of the most powerful government of the day. Napoleon's France between 1805 and 1814 towered over Europe, and for a time, it seemed that Napoleon's dreams of a reconstituted Roman Empire would be realized. Though Mathieu de Lesseps did not serve in the most sensitive posts, he was still young and well regarded by the Foreign Ministry. His brother Barthélemy was also a rising star in the diplomatic establishment, and was placed in more key positions than Mathieu. It was Barthélemy who represented Napoleon in St. Petersburg in the years before Bonaparte's disastrous Russian campaign, and who conducted the doomed negotiations in Moscow in 1812 as the general began his debilitating retreat. Graced with both a father and an uncle poised to occupy positions of influence in the emerging French Empire, Ferdinand inherited a predilection for the family business.

Mathieu was posted to Corfu for several years, but his wife and children did not accompany him. Instead, Ferdinand grew up on a palatial estate in the northern-Italian town of Pisa. Napoleon had conquered much of northern Italy before heading to Egypt, and Mathieu took advantage of French pre-eminence there to settle his family in style and comfort. Ferdinand grew up with siblings, servants, and tutors. Toward the end of his time in Pisa, he began to read French classics, such as the satirical plays of Molière and the works of Beaumarchais. Whether or not he had any interest in or aptitude for these studies is one of those unrecoverable details of the past. He was nearly ten years old when his family went back to Paris, after the Hundred Days, when Napoleon returned to France, rallied his supporters, and was crushed at Waterloo.

It would have been easy for Mathieu's career to collapse as well, but though he found the next years arduous, he got a diplomatic second wind and was posted to important positions: Philadelphia in the new United States, and Morocco on the coast of North Africa. Ferdinand, meanwhile, became a bourgeois young Parisian. Though he had been born in Versailles, it was the first time in his conscious life that he had actually lived in France. Napoleon had been replaced by the Bourbon Restoration, and France was suspended in a tense stasis. The forces of reaction wanted to revive the world as it was before 1789; the devotees of Napoleon looked for someone to carry on the empire; and the frustrated heirs of the Revolution tried to establish a republic of voters and citizens.

In the years immediately after 1815, while Lesseps was a student in Paris, the forces of reaction were ascendant. Louis XVIII, whose intellectual denseness was rivaled only by his physical obesity, pretended that his older brother, Louis XVI, had not lost his head under the guillotine but had only ceased to rule by some quirk of fate, which was now rectified by the restoration of his family to its rightful place on the throne. Both the Revolution and Napoleon were discredited, and the nobles who had fled or stayed in reduced circumstances tried to regain what they had lost. One of Ferdinand's classmates was the son of the Duc d'Orléans, who headed a branch of the royal family that had supported the execution of Louis XVI. In July 1830, the pretensions of Louis and his successor, Charles X, finally became too much for Paris, and the head of the house of Orléans, Louis-Philippe, replaced the deposed Charles.

Ferdinand's childhood acquaintance with royalty had little direct impact on his early career, but he was of that world. He was a member of what would later be called "the establishment," and from the time he could walk and speak, he had access to wealth and power. But his family was never in the upper tier, and Lesseps knew from a young age that he would have to make his own way if he was to enjoy the comforts and trappings that he had grown up with. His childhood created a template of ease and privilege, with ambition to maintain it.

Though Lesseps could not take his status in society for granted, it never seems to have occurred to him that he could lose it. That may be more a function of how he wanted to be seen than of how he actually experienced life. He left a voluminous trail of documents, especially concerning the Suez Canal, but as one of his biographers noted, it is hard to shake the sense that there's something missing. Lesseps published thousands of letters and several volumes of autobiography, yet through it all, he seemed preternaturally optimistic. There is never a moment of doubt, despair, or uncertainty committed to paper, or at least not committed to paper that Lesseps or his correspondents preserved. It is as if he went through life burning with energy and ambition, and never felt unsure; as if he existed in a constant state of clarity, and never once wondered whether his dreams would come to fruition or whether he would look back at his past with regret.[2]

When he turned in earnest to the Suez Canal, Lesseps was nearly fifty years old, and it may be that he had resolved any outstanding questions about who he was and what he should do with his energies. Most of the letters and memoirs that survive date from the latter part of Lesseps's long life, and they do not reflect the questions and concerns that may have bedeviled him as a young man still finding his way. The fact that Lesseps was not a reflective man makes it difficult to fill in the blanks of his first four decades. He did not have a philosophical streak, at least not on paper, and he was rarely introspective. In that, at least, he was quite ordinary. He may have accomplished extraordinary things, but, like most people of his day—like most people throughout time—he spent little energy looking inward. He was a Romantic in action, but not in his heart. He dreamed of fantasies that he somehow turned into realities, but he did not stop to ponder the eternal mysteries of life.

In that sense, he was the opposite of Enfantin. Whereas the Saint-Simonians were dedicated to unearthing the currents of history and progress, Lesseps strived to move history forward. The Saint-Simonians had a pragmatic streak that allowed them to finance and design railroads, bridges, and other engineering projects. But their material ambitions were always framed in a spiritual context. Lesseps was uninterested in metaphysics, and that may have given him a competitive advantage over those who were. All humans have the same twenty-four hours. Pursuing the ineluctable questions of who we are and why we are here takes time. Lesseps had unusual energy and drive. Perhaps, just perhaps, that was because his inner dialogue was inseparable from his external goals. Rarely stopping to ask why, he pressed ahead until some equal and opposite force stopped him.

His early career followed the same course as those of his father and his uncle. A calling as a diplomat suited a man of even temperament.

His father had benefited from the patronage of one of the most astute diplomats of the age, Talleyrand himself. Whether he received Talleyrand's wisdom directly, Ferdinand de Lesseps was certainly schooled by it. Talleyrand managed to survive and even thrive as successive regimes rose and fell, and he understood as well as anyone that a successful diplomat had to know the details of his environment and adapt accordingly. Having mastered those nuances, the diplomat also had to determine when to speak and when to be silent, when to be direct and when not to be. Duplicity, Talleyrand claimed, was never effective, but keeping one's tongue often was.[3]

Ferdinand appears to have followed these guidelines. He was posted to Lisbon in 1825, and that was the first of many years he spent in Portugal and Spain. Like Italy, the Iberian Peninsula was not a stable nation-state with inviolable borders. For more than a century, the French had been meddling in Spanish affairs. Wars had been fought, and the French had tried repeatedly to annex Spain or make it a subsidiary under a Bourbon king. In Iberia, Lesseps entered high society— not just as France's representative, but as the cousin of the Comtesse de Montijo, who was pregnant with a daughter who became an empress and who would later play a key role in the success of the canal. But that was years away, and before long Lesseps was posted elsewhere, to North Africa, and then to the scene of his father's success, Egypt.

While France was experiencing rapid changes and occasional turmoil, Lesseps lived thousands of miles away. He identified with France, but he spent a remarkably small portion of his life there. In some sense, he was more a citizen of nineteenth-century Europe than he was of any one country, and he embodied the virtues and vices common among elite Europeans of the era. He could converse in several languages. He wore the latest fashions, though he was not a dandy. He was meticulous in his physical presentation, eventually growing a mustache and sideburns in the style of the day. He knew how to dance and how to conduct himself at a ball or a formal dinner. Like the Saint-Simonians, he believed that never before in human history had any set of people been more graced with intelligence, civility, reason, and creativity than the inhabitants of Western Europe, and he felt certain that the world would be a better place when he died than it had been when he was born. He was confident to the point of arrogance. He sought acclaim because he thought he deserved it, and he was tireless in its pursuit.

For most of the 1830s, Lesseps lived in Egypt, dividing his time between Alexandria and Cairo, and he made regular trips back to Paris. In addition to being a commercial hub, Alexandria was where the pasha and his family spent summers, away from the stifling heat of Cairo. But Cairo remained the political, cultural, and religious center of Egypt, and when Mimaut left as consul in 1835, Lesseps was promoted to replace him. As consul, Lesseps pursued the same policies his father had. He nurtured the relationship with Muhammad Ali, who had become even more crucial to French goals internationally. For his part, the pasha, who was also known by the Turkish title *wali* (for "governor"), was not satisfied with commanding the Nile. He looked north to Constantinople and sensed that the Ottoman Empire was weakening. Though he paid nominal allegiance to the sultan, Muhammad Ali viewed matters in terms of power. The Ottomans were losing it, and he intended to gain it.

In 1833, his armies had almost toppled the sultan. Led by his elder son, Ibrahim, Egypt conquered what is now Syria and Lebanon and continued north, deep into Turkey, advancing close to Constantinople, until the Europeans rescued the sultan from almost certain defeat and compelled the Egyptians to retreat. The status quo, however, was too unstable to last long. The English had tethered their foreign policy to an independent though feeble Ottoman Empire that kept the Russians from gaining control of the sea-lanes separating the Black Sea from the Mediterranean. The French had other plans. Though the Napoleonic Wars had ended with France humbled, the system established by the Congress of Vienna in 1814–15 did not preclude competition and conflict abroad. For the rest of the nineteenth century, European powers fought their wars by proxy, and one arena was the eastern Mediterranean. In fact, the competition over the slowly crumbling Ottoman Empire was a centerpiece of European politics in the nineteenth century. The contest even had its own name: "The Eastern Question."[4]

For the rest of the 1830s, Muhammad Ali nursed a sense of grievance at Britain for snatching away his victory. He was allowed to annex Palestine and Syria, but he had been denied the larger prize. The French government fed his anxiety. Though Lesseps never explicitly urged him to challenge the peace brokered by the English in 1833, he did not dissuade him either. The pasha needed no prodding, but he knew that without the support of France he could never succeed. He could fight the sultan on his own, but he could not stand up to both the fading armies of the Ottomans and the increasing might of the British navy.

Muhammad Ali's fleet had been destroyed by the British in the 1820s in order to prevent the pasha from suppressing the Greek rebellion against the Ottomans. Muhammad Ali then built another fleet, less because he planned to use it against the British than because a navy was something great rulers possessed. He enlisted his thirteen-year-old son, Said, as an ensign in the nascent flotilla. He didn't expect the boy to do anything, but he did expect him to learn how to do something. The problem was that the young Prince Said was enormously fat. At thirteen, he weighed nearly two hundred pounds, and that was intolerable.

Most of Egypt was desert and had been ruled for centuries by Turkish lords. The government controlled the Nile Valley, its crops, its peasants, and its cities. But the desert was dominated by Bedouin tribes, who divided the sands into large, fluid fiefdoms. They took tribute from villages—though under whatever rubric, it was protection money pure and simple. The Bedouin shaikhs also struck a tacit agreement with the government in Cairo. The Bedouins were left alone, provided that they paid their minimal taxes and that their raids on caravans or on local villages did not get out of hand. The Bedouins profited from the camel trade south to Sudan and the Sahara, and from the pilgrimage route that went from Cairo across the Isthmus of Suez and down into the Arabian Peninsula to the holy cities of Mecca and Medina. The Turkish Mamelukes, meanwhile, carved up the country into tax farms. The arrangement bore some similarity to what was found in Europe until the early nineteenth century. Lords and religious leaders took profits from land tilled by peasants, who had little more freedom than slaves.

Both the ruling Turks and the Bedouins placed high value on warfare and dismissed corpulence as the effete weakness of a decadent court. Said loved to eat, but Muhammad Ali was determined that he lose weight. He ordered the young prince to undergo rigorous daily exercise, including rides in the desert and climbing ship masts. "I am disgusted at your weight," his father wrote. "It is in your power to lose that corpulence which offends all people and to acquire a slim body. . . . Your tutor has tried to conceal your condition and to protect and encourage your hateful corpulence." The pasha was losing his patience. He was trying to make Egypt European, and that meant

re-educating the elite of tomorrow. He worried that Said would be a liability, and that people would look at his son and lose confidence in the
father. Muhammad Ali made his feelings clear in another letter to Said:
"I will not permit people to humiliate me and say that I am incapable of
educating my sons when I educate the sons of all the others."[5]

Somewhat desperate at his inability to control Said's weight,
Muhammad Ali turned to an unlikely ally, the French consul. Though
the young prince's girth was hardly a state secret, this was an unusual
request to make of the representative of a foreign government. But the
pasha had known and trusted Lesseps's father, and the new consul had
obviously made a favorable impression. He was seen as discreet and
fair, and Prince Said was apparently fond of him. Implored by the
pasha to do something, Lesseps promised that he would try.

As it turned out, however, Lesseps did not have the heart to be
Said's dietician. Already, the prince's food intake was being carefully
policed by other members of the court, as well as by naval officers
nominally in command of the thirteen-year-old ensign. Said evidently
played an elaborate cat-and-mouse game with his various overseers.
He was by all accounts a sweet boy, but when it came to food, he was
capable of creative duplicity. Rather than challenge the prince at a
game at which he excelled, Lesseps chose the path of least resistance:
collaboration. He and Said shared a passion for horseback riding, and
he began to ride with the teenager almost daily. And rather than moni-
tor Said's intake, Lesseps broke every rule and gave him food.

Actually, he gave him macaroni. Said was especially fond of macaroni, and Lesseps became his provider of the fattening pasta. Every
now and then, the prince would drop a few pounds, but on the whole,
he remained as fat as ever. He became a skilled horseman, but never a
slim prince. In later years, the unwillingness of Lesseps to deprive a
hungry young boy of the food he craved would prove to be the most
valuable act of quiet defiance he ever committed. Some have made too
much of the story, and have claimed that a plate of pasta changed the
course of history. In any case, the exchange did cement a friendship
that had lasting consequences. Muhammad Ali never discovered that
the consul was deceiving him, or if he did, he ignored the matter. After
all, Said was his son, whatever the boy's weight.

Lesseps stayed in contact with Said when the prince was sent to
Paris to study. Though Lesseps spent most of his time in Egypt, he
periodically returned to France, and visited the prince each time. In

1836, on one of these trips, Ferdinand met a wealthy young woman named Agathe Delamalle. He courted her on his next visit, and they were married in December 1837. After a month and a half celebrating his honeymoon and setting up a household, he returned to Egypt. A son was born in November 1838, and Lesseps became close with his mother-in-law, Angélique, and brother-in-law, Victor, both of whom would remain trusted confidants for decades thereafter.

The next years in Egypt were dramatic. Muhammad Ali staged another invasion of Anatolia. He considered himself a legitimate contender for the Ottoman throne, and after years of thwarted ambitions, he was finally pushed to the breaking point by Sultan Mahmud. Allied with the British, Mahmud viewed the power of his supposed vassal with consternation. Muhammad Ali had almost occupied Constantinople in 1833, and the sultan's army was not much more equipped to resist him six years later. A commercial treaty between the sultan and the British threatened to cut off a vital source of Muhammad Ali's income while enhancing the rights of British merchants in Egypt. That was the casus belli. After several skirmishes, Muhammad Ali's army, commanded by his elder son, Ibrahim, annihilated the Ottomans at the battle of Nezib and advanced toward Constantinople in the fall of 1839. The sultan died before he learned of the defeat, and his young successor assumed power in unenviable circumstances.

Though he had the pasha's trust, Lesseps played only a minor role in the crisis. The important decisions were being made in Paris, London, St. Petersburg, and Constantinople. As had happened in 1833, the British scrambled to prevent Muhammad Ali from overthrowing the sultan and assuming control of the empire. The Russians supported the British, while the French lobbied for Egypt. None of them really knew what Muhammad Ali intended to do. The British believed that the pasha planned to overthrow the sultan and set himself up in his stead. But in conversations with European diplomats such as Lesseps, Muhammad Ali said that he merely wished to preserve his authority in Egypt and Syria, and that military force, though regrettable, was the only way he could make his case to a sultan who was unsympathetic and under the sway of a malevolent Great Britain and its foreign secretary, Lord Palmerston.

For Palmerston and the British government, the equation was simple: "Egyptian civilization," wrote Palmerston, "must come from Constantinople, and not from Paris, to be durable or consistent with British interests." That was not all. "France, in protecting Mehemet [sic] Ali," he continued in another letter, "means to establish a new second-rate maritime power in the Mediterranean, whose fleet might unite with that of France for the purpose of serving as a counterpoise to that of England."[6] These fears of Egypt as a naval power in cahoots with France were absurd. The Egyptian navy could patrol the coast near Alexandria, but it was no threat to England. However, Palmerston not only wanted to contain Egypt, he also wanted to keep the Ottomans weak and prevent France from challenging England in the eastern Mediterranean. If Muhammad Ali, or any Egyptian ruler, supplanted the sultan and allied with the French, these goals would be defeated. For the next two decades, Palmerston never wavered from his policy.

Under instructions from Paris, Lesseps did what he could to support the *wali*. He also engaged in negotiations with the other European consuls in Egypt. The crisis ended in 1840 with Muhammad Ali humbled once again, though at least he was able to salvage a greater degree of autonomy within Egypt. As a sop to his ambitions, he also secured an agreement from the new sultan that Ibrahim would inherit the governorship of Egypt when Muhammad Ali died. That was a significant concession, for sultans rarely replaced governors with family members. Muhammad Ali wanted to create an Egyptian dynasty, and though the European powers forced him to back down from a military confrontation, he did succeed in laying the foundation for future generations of his family to rule the Nile.

France had been humiliated by Palmerston, and Lesseps would never forget that, in Egypt and in Constantinople, Britain and France were adversaries. Napoleon's invasion had been the first indication that Egypt would be a theater of French and British competition, and the crises of 1833 and 1839 then formed the backdrop to the intricate diplomacy that characterized the building of the Suez Canal. The mutual suspicion between Britain and France over Egypt was shared by nearly all men of affairs on both sides of the English Channel, and when Lesseps first introduced the idea of the canal to the British public in the 1850s, he was greeted with skepticism and distrust.

But in 1840, all Lesseps knew was that the crisis had been settled and that French interests had not been served. He does not seem to have been deeply troubled. It was only one battle in what promised to be a long and uncertain war for European supremacy. It may have stung to leave Egypt on the heels of this defeat, but Lesseps did not feel personally responsible for the outcome. Events had been determined in Paris by his superiors, and those superiors opted to send him to a new post, one that he found extremely agreeable. He was appointed consul in Barcelona, where he was joined by his wife and his burgeoning family. As attached as he had become to Egypt, he was a professional diplomat. He could not advance his career by remaining in Egypt, and he looked forward to facing the challenges of a region torn not by great power rivalry, but by a nasty internecine struggle between republicans and royalists.

Lesseps spent the next eight years in Spain. He helped negotiate a temporary truce between the rivals, though the problems of Barcelona would continue long after he left and well into the twentieth century. He was comfortable in Spain, having spent considerable time there as a young man, and he had Spanish relatives who welcomed him. His family grew, and his marriage was warm and loving. That was not a given. Many marriages were procreative and economic alliances, and though romantic love was beginning its ascent, it was still neither expected nor necessarily desired. The love between Ferdinand and Agathe was a bonus for them both, as were the tight bonds that developed between Lesseps and his in-laws. In Spain, he matured from a respected young diplomat to a man of stature within the Foreign Ministry, and when the tumult of 1848 suddenly ended the tepid reign of Louis-Philippe, Lesseps was recalled to Paris, not because his star was falling but because it was on the rise.

The revolutions of 1848 affected not just France but all of Europe. In the capitals and major towns of Italy, Austria, Hungary, and Prussia, people took to the streets and demanded change. Their tactics were sometimes planned, and at other times only the reflection of inchoate rage. The French uprising was sparked by a faltering economy, high levels of unemployment, and a general sense that the regime of Louis-Philippe was incapable of addressing the nation's problems. The years between the revolutions of 1830 and 1848, far from healing the country, had hardened the divisions between those who wanted a strong monarchy, those who wanted a constitutional monarchy, those who yearned for a republic governed by the elite, those who demanded a democratic society of universal suffrage, and those who hoped for a socialist society modeled on the radical days of the early 1790s.

Louis-Philippe's government collapsed in February, and for the next months, Paris was in a political frenzy. Through the spring, it seemed as though a rough new order—part republican, part conservative—would be worked out. But revolution spread throughout Europe, and the more conservative forces in France looked nervously at what was going on abroad and decided to halt the brief efflorescence of social innovation. General Cavaignac brutally suppressed street demonstrations in Paris at the loss of thousands of lives, but he had the support of fearful conservatives and the provinces. That did not mean he had a mandate to establish a military dictatorship, however, and elections were scheduled for the year's end. Meanwhile, throughout Europe, the revolutions were defeated, making 1848, in the words of one historian, "the turning point at which modern history failed to turn."[7] Witnessing these defeats, a young Karl Marx so despaired of change that he developed a utopian philosophy that would eventually have considerable success, before it too collapsed. The revolutions of 1848 may have failed in their time, but they led to even more significant uprisings decades later.

For Ferdinand de Lesseps, the events in France led to a posting as minister plenipotentiary in Madrid. As ambassador with extensive powers, Lesseps was supposed to help keep the peace in Madrid and prevent, if he could, similar revolutions from overthrowing the Spanish royal family. There was, in fact, a revolt in Madrid, but it fizzled before it could seriously threaten the monarchy, and Lesseps was able to call his mission a success. During these months, he also developed a warm relationship with his young cousin Eugénie de Montijo.

But the situation was rapidly shifting in Paris, and the rising fortunes of Napoleon Bonaparte's grandson Louis-Napoleon led to Lesseps's recall. Louis-Napoleon had run for the new office of president at the end of 1848, and had, much to everyone's surprise but his own, won by an overwhelming margin. Doling out rewards to his supporters, he made his brother the new ambassador in Madrid, and that meant that another position had to be found for Lesseps. He was on his way to take up a post in Berne when he was suddenly redirected to Rome. Berne would have been dull but safe. Rome was to be his downfall.

# CHAPTER FIVE

# EGYPT AND ROME

A T THE SAME time that Lesseps was heading into a trap, the pasha of Egypt was descending into senility. Nearly eighty years old, Muhammad Ali clung to power until he no longer realized that he possessed it. By the middle of the 1840s, his son and the general of his army, Ibrahim, had assumed day-to-day responsibilities of governing the country. By 1848, Muhammad Ali could barely carry on a coherent conversation, and though Ibrahim died in the spring of 1849, the pasha registered no awareness of the fact. He lasted several more months and then quietly passed away.

He left a vastly different Egypt from the one he had first glimpsed in 1801, and his role in determining the shape of modern Egyptian history is impossible to overstate. Egypt may have been at a strategic cross-roads for the Europeans, and it almost certainly would have come under pressure from France, Britain, and other powers, later if not sooner. But Muhammad Ali set a particular course, one that simultaneously embraced European culture and rejected the idea of European superiority. Early on in his career, he realized that the French and the British had better armies, better organization, and better education than could be found anywhere in the Ottoman Empire. And early on, he decided that the only way to keep the French and the British from overwhelming the Ottoman Empire was to adopt their methods.

Easy enough in theory, this proved a challenge in practice. Muhammad Ali was hardly the only non-European to come to this conclusion in the nineteenth century, but he was one of the first to see which way the winds of history were blowing. Instead of blindly resisting the changes heralded by the rise of European power, he elected to bend and to borrow. That he didn't succeed in remaking Egypt quickly enough to prevent the European incursion is not surprising. That he came so close is extraordinary.

His military campaigns, which so alarmed the English in the 1830s, were only possible because of the reforms he had instituted in the decades before. It was not just a matter of reorganizing the government in order to generate more revenues, though that was vital. With one foot in the old world and one in the new, Muhammad Ali tried to transform Egypt into a country capable of competing with the more technologically advanced Europeans, without unleashing the forces of revolution. He admired the French military and their bureaucracy, but not their ideology. He built an Egyptian army of tough, disciplined soldiers who learned the latest stratagems but were not encouraged to think for themselves. As Enfantin had discovered, Muhammad Ali was an absolute monarch. If the pasha did not wish something to be done, it would not be done. He allowed no one to challenge his will, and he used the time-honored methods of kings and tyrants to suppress dissent: violence and the threat of it.

What made him different was an unusual degree of intellectual curiosity and a willingness to experiment with new ideas. Much like the first Hanover king, George I, who spoke only German yet ruled an English nation, Muhammad Ali spoke Turkish and Albanian but no Arabic in a country that consisted mainly of Arabs. Though he never mastered Arabic, he did learn to read and to write, and he commissioned translations of the works of Voltaire, Montesquieu, and Machiavelli. He also studied the careers of Alexander the Great, Julius Caeser, and Napoleon, looking for clues to successful empire-building.

Recognizing that Europe possessed knowledge that Egypt could use, he sent hundreds of young Egyptians abroad in the 1820s and 1830s to study in France, England, and Italy, with the largest contingent in Paris. They were selected for their intelligence and curiosity, but Muhammad Ali made it clear that he did not want them to become too fond of Europe. Though they were to learn new languages and digest innovative ideas, they were not to apply them to life in Egypt. Only those ideas that enhanced his power and did not threaten his autocracy were acceptable. As he instructed one student who returned from Paris offering his services to reorganize the administration in some provinces, "It is I who govern. Go to Cairo and translate military works."[1]

And it *was* he who governed. After he appropriated the lands of the Mamelukes that he had slaughtered in 1811, he doled out parcels to his retainers and family members. But it remained his land, and he could revoke those grants whenever he wished. Though the years before Napoleon's invasion had been a period of decline, for most of its history Egypt was a centralized state. Muhammad Ali took that centralization one step further. Traditionally, Egyptian rulers collected revenue from the farmers along the Nile and from the merchants in the cities and then left daily life alone. Muhammad Ali changed that, and took an aggressive role in determining what crops were grown, by whom, and where. He also dictated to merchants what they should sell and what they should buy. It would have been less complicated if he simply left the system that had prevailed for millennia intact. But he could not, for one simple reason: he needed more income.

In Egypt as it was before Muhammad Ali, there had been a strict limit to how much revenue any ruler could extract. The economy of Egypt in the early nineteenth century remained much as it had been for thousands of years. The country was mostly barren, and less than 5 percent of the land supported the population of three or four million people. The other 95 percent of the country was desert, ranging from the dunes of the Sahara in the south and west to the high desert of Suez and Sinai in the east. Christian hermits and camel-trading Bedouins made these deserts home, but they were few and far between.

In the villages scattered along the banks of the Nile, from the cataracts of Aswan in the Nubian south to the alluvial delta where the Nile opened to the Mediterranean in the north, there had been few changes over the centuries. Egypt had been invaded many times, but rarely had armies penetrated beyond Cairo or Alexandria. Village life revolved around a local headman, or shaikh, and religious life around the Sufi lodges that dotted the countryside. The tombs of holy men became shrines, places where locals gathered to pray for fertility, both for their crops and for their daughters. Religion was not a dogmatic force, though it wasn't progressive either. The mosque was a place of worship, but it was no more important in people's existence than the local church was in eighteenth-century France. In fact, the only appreciable difference between the daily life of Europe and Egypt until the nineteenth century was the temperature, the food, and the clothing. In every other respect, most parts of the world consisted of landed peasants growing crops, getting married, cleaving to parochial superstitions, having children, and growing old.

True, the color of Egypt was different from that of Europe. People wore long robes, rode camels, smoked water pipes in Cairene coffee shops, chanted the Koran in Arabic, and prayed to the same God of Jesus and Moses in a different way. Snake charmers charmed snakes, merchants of the bazaar grew rich from the trade of silks and spices from the East, and crowded alleys in city streets wrapped around each other in impenetrable mazes. Slaves were sold in open-air markets, and weddings paraded through the streets accompanied by a cacophony of reed pipes and drums. Palm trees swayed along the banks of the Nile, and small boats plied its currents. And here and there, an ancient mound of sand and stone in pyramid shape reminded the inhabitants that people had lived in the Nile Valley far longer than anyone could remember.

These picturesque qualities of Egypt were all that most Europeans noticed. They saw differences and overlooked similarities. Centuries earlier, when travelers such as Marco Polo reported back, Europeans had admired what they saw in the cultures of the East, but by the eighteenth and nineteenth centuries, the attitude was often one of disdain. When Napoleon invaded, the French believed that Egyptian society was backward and in need of improvement. But Egyptian elites such as Jabarti regarded the French as rude and uncivilized, and they sometimes saw the French more clearly than the French saw themselves. When Europeans examined the cultures of the world in the nineteenth century, they avoided looking themselves in the mirror. When they encountered the streets of Cairo or Calcutta, they somehow forgot about the chaos of Paris and London. When they saw petty brutalities carried out by local potentates, they condemned the backwardness while conveniently eliding their own legacy of violence and autocracy. When they saw religious dogmatism, when they noted widespread illiteracy, they depicted these as symbols of Eastern regression, even as multitudes of their own countrymen suffered from the same problems.

Over the next century, European attitudes toward Egypt, toward the Ottomans, toward India, China, and the rest of the non-Western world became more complicated, and different prejudices developed for different regions. But most Europeans never strayed from the essential belief in the inferiority of the East. Elites of the East, on the other hand, despaired as their attempts to resist Western encroachments failed. Soon, just as the West dominated the East militarily and economically, it dominated culturally as well. Eastern societies, said the Europeans, were weak because they rewarded obedience instead of initiative, tradition instead of progress, and religion instead of science. To the victor, the spoils; and one of the spoils is the power to interpret events. This Western perspective of Eastern weakness and decadence has been both defended and excoriated over the years. It has been derided as "orientalism," as a Western inability to grasp the sophistication of Eastern cultures.[2] But at the time, the states of the West were expanding unchecked throughout the globe. They had no need to grapple with the cultural complexities of their adversaries. The same was not true on the other side.

Until the middle of the nineteenth century, Egyptians and Ottomans still had hopes of resisting the West and even challenging the Western powers. They could only do that, however, by learning from them. The trick was to borrow enough but not too much. Muhammad Ali respected Europe because it was strong, but he also saw himself as part of a world with its own traditions, mores, and culture. He was a member of the Ottoman mercenary class, which had a proud history. He was now part of the Egyptian ruling class, and as such, he had inherited a legacy that predated the upstarts of Europe. He believed that it was possible to redress the imbalance between East and West without sacrificing the essence of his own culture. To hold his own against Europe, he was willing to put his country through a difficult period of reform. There was only one thing he lacked: money.

To rectify that, he ordered the peasants, the fellahin, to grow specific crops. Rather than just growing the grains and fruits that had sustained the country and provided some food for export, he committed vast acreage to tobacco and a new strain of cotton. The cotton was a long-fiber variation called Jumel, after the Frenchman who invented the strain. By the late 1820s, Jumel amounted to as much as a quarter of the state revenues. In order to increase his income, Muhammad Ali had to tame nature. The Nile floods guaranteed regular inundation of the fields, but they were unpredictable and frequently destructive. It also took countless hours to construct the small canals and ditches that could transport the water to fields not directly on the riverbanks. Without some way to regulate the floods, Muhammad Ali would never be able to predict his revenue year to year, and that would make long-term planning difficult. So he began an aggressive campaign of building canals and barrages, and turned to Europeans, to the Saint-Simonians and to Linant de Bellefonds, to help with the engineering.[3]

The quest for money animated not just Muhammad Ali but his

heirs. That insatiable need led to changes in Egyptian society, and it eventually led his successors to grant a concession for the Suez Canal and then go into debt in order to fund it. Without money, Muhammad Ali and his heirs could not purchase new weapons, build new schools, construct canals and railroads, pay for the bureaucracy. Without money, they could not re-create Egypt, and if they could not re-create Egypt, they knew that, sooner or later, the Europeans would do what great powers have always done: they would invade and conquer.

Though he rejected European political ideas, with each passing year Muhammad Ali became a more avid admirer of European power. That was also true of much of the Egyptian ruling class. The students who were dispatched to Europe wrote of their experiences with a mixture of admiration and revulsion, fascination and disgust. In 1826, a young shaikh named Rifa'ah al-Tahtawi went to Paris at the age of twenty-five. Though Muhammad Ali had issued strict instructions that the students were not to mix too freely in French society, his ability to control their activities was limited, and they inevitably went out and about. Tahtawi learned French, studied Napoleon, read Rousseau, and was a frequent dinner guest at the salons of leading writers and intellectuals such as the orientalist scholar Sylvestre de Sacy. For all his immersion in French culture, however, Tahtawi remained skeptical of liberalism and suspicious of democracy. He also found the French a bit strange.

He noticed that in Paris the streets were filled with a wide array of vehicles to carry goods and people. He couldn't understand the need for such variety, and he was also unimpressed by the cleanliness of French streets. Paris in these years still had open-air graveyards near what is now Les Halles, and disease was common. Tahtawi did admire the French press; newspapers were unknown in Egypt, and the concept of people airing different views about politics he found both startling and exhilarating. And then there was the French postal service, which was so sacrosanct, he observed, that "letters exchanged between friends, colleagues and above all, between lovers abound, since everyone knows that his letter will be opened by no one but the addressee. When a lover declares his love to his beloved, he does so by correspondence. It is also by correspondence that they make their rendez-vous."[4]

It was an innocuous observation, but the postal service was more than a convenient way to arrange lovers' trysts. It was an integral component of Europe's global expansion. Empires need to communicate with their far-flung outposts, and the modern mail was the way that the French and the British communicated with theirs. European empires of the nineteenth century were commercial behemoths, and merchants in Marseille and Manchester had to be able to contact their agents and counterparts in Alexandria and Bombay. The Mediterranean routes were well established, but it took many months for information and goods to flow between England and India. Even with the first steam-powered ships in the 1820s, it took at least 113 days to travel the six thousand miles from London to Calcutta around the Cape of Good Hope.

The best way to shorten that distance was to take the mail across the Mediterranean to Alexandria, and from there down the Nile to Cairo, then across the desert separating Cairo and the port of Suez, and then from Suez by steamer to India. One enterprising young English naval officer, Lieutenant Thomas Waghorn, knew a good business opportunity when he saw it, and after serving in the Napoleonic Wars and in various positions with the East India Company, he dedicated his life to organizing the Alexandria-to-Suez route. That was not a simple task. The port of Suez, for instance, was deep enough for large ships, but if it was to act as a port for steamers, Waghorn had to haul large quantities of coal there and create storage systems for it. It took years of careful politicking, in both Egypt and England. By the late 1830s, Waghorn's overland route finally graduated from a dream to a fact, and he became the official agent in charge of transporting mail from Alexandria to the port of Suez.

The Waghorn route, though primarily used by the postal service, also catered to passengers, though not with creature comforts. The journey entailed twenty-four hours of bumpy travel to cross the desert, and it was usually brutally hot. The lack of comforts had less to do with Waghorn than with the excessive cost of transforming the roads into more than outlines, and upgrading the steamers along the Nile. Instead, travelers were treated to a premodern version of planes, trains, and automobiles, complete with ornery camels, donkeys, and only intermittently cooperative Bedouins. In addition, though Muhammad Ali technically approved of the Waghorn system, the Englishman was forced to negotiate with the local Bedouins for safe passage across the desert from Cairo to Suez, and as with all protection rackets, this one needed frequent renegotiation.

Waghorn faced competition in the 1840s from other entrepreneurs, who built hotels in Alexandria and Cairo as well as guesthouses along the eighty-four-mile desert route. These competitors aimed at the passenger traffic, while Waghorn remained in control of the postal routes. He also had a stake in the railroad that was being constructed between Alexandria and Cairo, which would of course help his business and shorten and simplify the route. But by creating a system of regular transportation, Waghorn set a precedent. Lesseps, who was still consul when Waghorn's route began, claimed in later years that Waghorn had shown the world that a shorter route between Europe and the Orient was both feasible and desirable. In 1870, to honor Waghorn's memory and to placate English pride, Lesseps commissioned a statue of the Englishman to be placed near the canal. "He opened the route, and we followed him," Lesseps said at the dedication ceremony. "When English navigators pass by this monument to Waghorn, erected by the French, they will remember the intimate alliance which should always exist between the two nations that have been placed at the head of world civilization, not in order to ravage the world, but in order to enlighten it and bring it peace."[5]

Though Muhammad Ali had vetoed the idea of a direct canal, the popularity of the Waghorn route kept the idea of a Suez Canal alive, and Enfantin soon re-entered the scene. Though the Father had moved on to other projects, he had lost none of his messianic zeal, and he had also acquired the patina of a successful entrepreneur as a result of his involvement in the development of railroad lines in France. His new wealth fueled his old ambitions, and in the 1840s, after agitating for French schemes in colonized Algeria, he again turned his energies toward the creation of a canal. In the interim, the field of competitors had grown.

By the 1840s, Egypt was teeming with European businessmen and diplomats. Muhammad Ali's modernization programs acted as a beacon for thousands of Europeans looking to make a fortune and a reputation in an untapped market. The prospect of a canal meant the possibility of trade of far greater magnitude. The overland route remained costly and time-consuming. Letters and passengers could be transported, but not bulkier, heavier items. Railroads were a distinct possibility, but water transport would be even cheaper, and faster. Along with Frenchmen such as Enfantin, Austrian, Italian, Dutch, Greek, English, Scottish, and German diplomats and merchants began to look more seriously at a canal project.

Like Leibniz before them, a group of Leipzig merchants petitioned
Muhammad Ali to allow for a preliminary canal project in 1845. He
rebuffed them, but not as forcefully as he had dismissed Enfantin a
decade earlier. He said that they should first sound out the great powers of Europe, and if England and the rest agreed that a canal was a
good idea, then he would think about it.[6] Weakened by the onset of old
age and by his bitter defeat at the hands of Palmerston, Muhammad Ali
accepted that Egypt's fate was inextricably linked to European politics.
Rather than ruling one way or another on the merits of a canal, he
deferred a decision and told the interested parties that the final verdict
would come not from Cairo but from the capitals of Europe.

Perhaps perceiving that the climate had changed, Prosper Enfantin
revived his dormant ambitions. Using his wealth and connections, he
assembled a group of engineers and entrepreneurs and raised 150,000
francs. At the end of 1846, in his apartment in Paris, Enfantin established the Société d'Études du Canal de Suez (Suez Canal Study
Group), to conduct further research into the feasibility of a canal linking the Red Sea and the Mediterranean. Enfantin now recognized that
the canal would have to be a multinational effort, and the society
included prominent engineers from England, France, and the German
states. Representing England was Robert Stephenson, son of the
famed inventor of the railroad and a figure of substance in his own
right; from France, Paulin Talabot, an adherent of the Saint-
Simonians; and from Austria, Reynaud de Negrelli. Dozens of others
(including two of Talabot's brothers) were involved, but these three,
plus Enfantin and the Saint-Simonian François Arlès-Dufour, were the
central figures.

The Study Group sent its engineers to Egypt in 1847. They coordi-
nated their plans with Linant de Bellefonds, and his assistance was
vital. When Muhammad Ali learned of the Study Group, however, he
reacted strongly. Age had made him suspicious to the point of paranoia.
He remembered, or one of his ministers reminded him of, Enfantin's
earlier attempts to obtain a canal concession, but because Enfantin did
not approach him directly this time, the pasha assumed that the Study
Group was part of some plot by the powers of Europe to undermine
Egyptian autonomy. Linant, who had served Egypt loyally as director of
public works, assured the pasha that the group was what it appeared to
be, and that, although a canal might still be a bad idea, it was at least
worth further study.

The teams dispersed to their designated regions along the canal. Stephenson, assigned to survey the area around the port of Suez, was the most perfunctory. Owing to his association with railroad developers, he soon lost interest in a canal and became an advocate for rail links between Alexandria and Suez. Representing Austrian interests, Negrelli believed that a canal could make both him personally and the Austrian Empire rich. Goods passing from the Orient through the canal on their way to Central Europe could transform Austria's Adriatic port of Trieste into a major entrepôt. Stephenson made a desultory pass through the south; Negrelli assiduously surveyed the lands and the coasts around the Bay of Pelusium. That left the interior of the isthmus to Talabot.

The brothers Talabot came from Limoges, and both were immersed in the nexus of business, politics, and ideas that made mid-nineteenth-century France such a churning, chaotic place. In tune with the times and in line with the Saint-Simonians, Paulin saw grand public-works projects not just as puzzles to be solved, but as integral components of the harmonious development of civilization. Like Enfantin, he hitched his career to the development of railroads, and he later was a central figure in the evolution of the French banking system. After trekking across the desert to the Bitter Lakes, at the midpoint of the isthmus, he spent weeks measuring and surveying. He came to a startling conclusion. The survey conducted almost fifty years before by Le Père was wrong. There was no meaningful difference in the levels of the two seas, and thus there was no potential problems of flooding. That was what Linant had been claiming all along, but he had never managed to convince Muhammad Ali to spend the money on a new survey. Talabot was blessed with the latest equipment and ample funds, and after checking and rechecking his data, he announced that a direct route linking the two seas presented no technical obstacles.

Even so, Talabot did not recommend a direct route. In his final report, he said that, though it was feasible, it was not desirable. It would require the construction of a new port in the north, and it faced the difficulty of no fresh water, no supply lines, and no proximate labor. Instead, he suggested the old idea of an indirect link. Personal rivalries shaped this recommendation. Talabot was allied with Enfantin, and Enfantin had a falling-out with Linant. Advocating a direct route would have meant deferring to Linant, who knew the terrain better than any man alive, had long been on record as a defender of the direct canal, and had the confidence of the pasha. Enfantin, however, was not a team player. Whatever the technical merits of a canal cutting the isth-mus, taking a back seat just wasn't his style, and as went Enfantin, so went Talabot.[7]

For the rest of 1847 and into 1848, the Study Group was active. Plans were drawn and redrawn, and letters passed freely among the capitals of Europe as Enfantin, Talabot, Negrelli, and Arlès-Dufour tried to win friends and influence people. The resistence of the English government and the passivity of the pasha, however, kept the idea from making the transition from a scheme to a project. It was much dis-cussed, but nothing actually happened. Or, at any rate, nothing that was apparent at the time. There was, however, one letter exchanged, between members of the *société* and the French consul in Spain, Fer-dinand de Lesseps, who was enthusiastic but immersed in a promising diplomatic career far from Egypt.

The future of the canal darkened even more when Muhammad Ali died. He had been an opponent, but he had hinted that if the powers of Europe wished it he would be open to persuasion. The same could not be said for his grandson and successor, Abbas, who became viceroy when the pasha finally died in the autumn of 1849.

Though Said was Muhammad Ali's son, the laws of succession dic-tated that power be passed to the oldest male descendant, and that was Abbas. Abbas is a murky figure, and most of what is known about him comes from his opponents. Yet even neutral chroniclers, both since and at the time, searched unsuccessfully for good things to say about him. He was portrayed as cruel and reactionary. One Egyptian histo-rian of the late nineteenth century described Abbas as a spoiled child who could be both irritable and threatening. He had been orphaned at a young age, and his education was overseen by Muhammad Ali. As harsh as he had been toward the pudgy Said, the pasha was no gen-tler with Abbas, who was at best lazy and at worst stupid, or so most people said. When Abbas was a teen, the pasha had dismissed him, saying, "He is incapable of work and shows proof of a crude nature." Once in power, Abbas wreaked the revenge of a scorned child and rejected the modernizing policies of Muhammad Ali as misguided and dangerous.[8]

In decree after decree, he tried to dismantle what his grandfather had so meticulously constructed. Abbas placed restrictions on Euro-pean merchants, and refused to meet with their consuls. He recalled the Egyptian students studying abroad, shut the schools that taught science, and made it known that Egypt was now closed to business. He did support the completion of the Cairo-to-Alexandria railroad under the auspices of Robert Stephenson, probably because that would allow him to consolidate his control over the Nile Delta without giving up any Egyptian sovereignty. Though Abbas was widely disliked at the time and has been reviled since, he had good reason to distrust modernization. By midcentury, Egypt had become more porous to European influence, but no more able to resist European imperialism. Abbas and his circle may have reacted blindly to the threat, but they gauged it correctly. With the process begun by Muhammad Ali only half complete, Egypt was more vulnerable. The old structures of power had been so weakened that they could no longer effectively resist outside influence, even passively, and new ones weren't sufficiently in place. Egypt was left betwixt and between, neither modernized enough to achieve Muhammad Ali's dream nor isolated enough to allow Abbas to retreat from the world without consequences.

With Abbas in control, there was no chance for Enfantin and the Study Group to implement their plans. He refused to meet with them. He exiled, arrested, or executed many of the reform-minded ministers who had been in Muhammad Ali's employ, and he spent his few years in power trying to change the order of succession so that his son and not his uncle Said would inherit the mantle.

If Abbas appeared to doom the Suez Canal, then the events in Rome in 1849 seemed likely to doom the man who would in time became the canal's creator. Lesseps was sent to Rome to bring peace to a city that was besieged by multiple armies, with Austrian troops on one side and French infantry on the other. Between them were the forces of the newly declared Roman Republic, led by Mazzini in an uneasy alliance with Garibaldi. The Republic was dedicated to the principles of liberalism that flourished in 1848, and its leaders pressured the Pope to renounce his political power. The Austrians wanted to crush the Republic and restore papal rule. The army of France, meanwhile, answered to the newly elected government of President Louis-Napoleon, and he was unsure whether his interests lay with the Pope or with the Republic.

To make matters worse, General Oudinot of the French expeditionary force intended to resolve the situation by force. He wanted to defeat the Republicans and then eliminate the Austrians as the protectors of the papacy. Lesseps was sent to Rome with minimal instructions. He was told "to deliver the States of the Church from the anarchy which prevails in them, and to ensure that the re-establishment of a regular power is not in the future darkened, not to say imperiled, by reactionary fury." Other instructions dictated that he do everything he could "to secure genuine and real guarantees of liberty for the Roman States."[9] What that meant in practice was anybody's guess. Restore the Pope but force him to work in concert with Mazzini and the Republicans? Fight the Austrians? Supersede the French army and overrule General Oudinot? Lesseps could not have known, and he did not ask.

Arriving in Rome in early May, Lesseps found an intransigent Oudinot preparing to invade the city. He spent countless hours shuttling between the general and the leaders of the Republic, and though his relationship with all parties was strained, he eventually hammered out a provisional treaty in late May that allowed for both Roman self-determination and the temporary occupation of the city by French troops. With this compromise in hand, Lesseps decided that it was best to meet personally with the foreign minister. He returned to Paris in a hurried twenty-four-hour journey, not suspecting that his work was about to be repudiated and his good name ruined.

In Paris, he learned that the foreign minister had been replaced, by Alexis de Tocqueville, and that Oudinot had received authorization to invade. Lesseps was left with a useless piece of paper, and an unresponsive Foreign Ministry. Tocqueville may have lauded the democracy of the United States, but that didn't preclude him from the duplicity of high politics. Though he did not exactly betray Lesseps, Tocqueville did little to prevent him from being sacrificed. The Assembly took only a few days before declaring that Lesseps had exceeded his instructions and that the convention he had negotiated was null and void.

It didn't stop there. Louis-Napoleon needed someone to take the blame. Lesseps, a career diplomat who had been out of the country for most of the past two decades, was a relatively easy target. At least Ferdinand saw what was coming. He wrote to his wife, "It will not be the first time I have been alone against the world and have won through. . . . I well know how to ride out a storm." He then assured her that he was "never so calm as at the very center of troubles." He had a clean conscience, had done nothing wrong, and was sure that one day he would be vindicated.[10]

But worse days were ahead. Private remonstrances were not sufficient. He was brought in front of the Council of State in a mockery of a fair trial, and assailed for negotiating an agreement that compromised the honor and integrity of the French army and which undermined the dignity of France. He was then officially censured. Though not explicitly ordered to end his association with the Foreign Ministry, he was left in a professionally untenable position, without patronage or hope of a subsequent assignment. He resigned.

Decades later, Lesseps wrote a defense of his Roman debacle, and concluded that it had all turned out for the best: "The vote of blame was a fortunate one for me, as, returning to private life, I have since been absolved by my country."[11] That was easy enough to say after the triumph of the Suez Canal and the impending promise of the Panama Canal had catapulted Lesseps to fame and fortune. But in the summer of 1849, he could not have been so sanguine about his future. However, as his letters attest, he did have a wife whom he loved and who loved him; an eldest son, Charles, whom he was particularly fond of; an ally in his brother-in-law; two younger children; and a close bond with his mother-in-law. Agathe was stalwart in her support of her beleaguered spouse, and Madame Delamalle was more than simply sympathetic. She was rich.

Ferdinand was forty-four years old and unemployed. He had some family money, but he had no salary and no standing. Responding to his need, Madame Delamalle appointed Lesseps her agent for a large estate in central France, near Berry, and he convinced her to pay for renovations to the main house. He then dedicated his energies to turning the estate into a model farm, and he plunged into that work with the same intensity that he would later bring to more grandiose projects.

History then upended his life once again. In the summer of 1853, his son Charles became very sick, probably with scarlet fever. Agathe cared for him, and nudged him toward health. But she caught his fever, and was dead within days. One of her other sons, named Ferdinand after his father, died as well. Within the space of a few years, Lesseps had lost a career, a wife, and a child. He was left with two sons, a farm in central France, and a caring set of in-laws. He also had a deep well of unused energy, which was now augmented by a stream of grief. "I pass my life in the middle of cows, pigs, and lambs," he wrote in his diary. A solitary life as a yeoman farmer was not a long-term option, yet nothing else had presented itself.[12]

A canal might have been built without Ferdinand de Lesseps, but not then, and not for many decades. Though the idea had been floating through Europe, no one had been able to solve the Gordian knot formed by political rivalries, competing agendas, economic challenges, and engineering disputes. Lesseps himself would never have devoted himself so single-mindedly to implementing the project had he not suffered loss. In that sense, he was right to look back on this period as a blessing in disguise. Disgrace and the death of his wife propelled him. His pain became the source of his inspiration, and his ambition a way to heal himself.

This man in search of a destiny now received a gift from fate. In July 1854, a postman appeared in the courtyard of an old manor house Lesseps was restoring. The letter was handed to him by one of the workmen. Abbas had been assassinated, and Lesseps knew instantly what that meant.[13] Power had passed to the eldest male of the house of Muhammad Ali, to a thirty-two-year-old man with a fondness for all things French and a penchant for large plates of pasta.

# CHAPTER SIX

# A Journey in the Desert

A BBAS WAS MURDERED by two of his slaves, under mysterious circumstances that few had a vested interest in probing. Said assumed the reins of government with little resistance, and within weeks, the word was out: Egypt was again open for business.

Where Abbas had turned the country inward, Said looked to Europe. Where Abbas had tried to reverse the course set by his grandfather, Said intended to honor Muhammad Ali's legacy. Abbas had made one small concession to Westernization: he granted a group of British businessmen a concession to build a railroad from Alexandria to Cairo. But whereas Abbas had been stingy with such grants, Said was profligate. In almost everything he did, Said went to excess. His appetite for food was a corporeal reflection of his appetite for life. Had he been just a bit more restrained, or had he gathered advisers who encouraged his more spartan instincts, he might have steered a perfect course between opening Egypt to the commerce of the world and guarding Egypt against those who would exploit it. Unfortunately, there was no one cautioning against that next grant, that next party, or that next plate of macaroni. His heart was good, but he didn't know when to stop.

Of course, that only became clear years later, after his projects began to undermine the Egyptian financial system. His successor, Ismail, did little to reverse course. Said's decisions helped lead Egypt into a lethal embrace with a more powerful, rapacious Europe, but in his lifetime, he seemed to be making the right choices and doing all that he could to strengthen his country and keep it independent.

His teenage stint in the navy was only one element of a regimented education designed by Muhammad Ali. Part of Said's schooling was traditional. He was a young Turkish prince, the son of a beautiful, tall Circassian mother who was one of Muhammad Ali's concubines. His first years were spent in the harem among the female members of his father's court. Then he was placed in the care of male guardians and trained in martial skills such as riding, shooting, and fencing. He was tutored about the Koran, the life of the Prophet, and the golden age of the Arab caliphate in Damascus and Baghdad, and he learned about the great victories of the Ottomans and their glory days under Suleiman the Magnificent, who conquered Egypt early in the sixteenth century and nearly became the lord of Central Europe. Said loved the desert, and military exercises. He enjoyed wearing uniforms and watching his troops parade in front of him. One of his favorite pastimes was taking special guests into the desert, erecting a temporary court of elaborate tents, serving sumptuous meals, and spending several days watching military maneuvers.

But Said also received an education unlike that of the average Turkish prince. He had a French tutor named Koenig Bey and learned fluent French. He read the French classics and acquired a zest for French culture that was only intensified by his teenage friendship with Lesseps. He was sent to Paris in the late 1830s, where he passed many days roaming the Latin Quarter. His affinity for Paris never faded, and this was one of the reasons why he responded so favorably to the French when he assumed power.

Francophile that he was, he renovated the palaces made musty by the austerity of Abbas and filled them with furniture built in the ateliers of Paris. He purchased silks and damasks from Lyon and Marseille, and he collected ceramics from Limoges. In fact, during his reign, the Egyptian royalty acted and dressed much like European royalty. The Ras al-Tin ("Cape of Figs") Palace in Alexandria, where Said passed his summers, was redesigned to mimic a comparable royal abode in Vienna or Milan. And in his opposition to political reform, he was no different from the Austro-Hungarian emperor.

Said may have been a genial, generous soul with a penchant for luxury, but he was also an autocrat. In fact, it never seems to have occurred to him that there was any other form of government. He was the ruler, and his word was law. Technically, his authority to order executions was constrained by his legal status as a vassal of the Ottoman sultan. In practice, however, he had the power of life and death over millions of Egyptians. He could command twenty thousand peasants to travel hundreds of miles to work on whatever projects he wished. He could spend the money that flowed into his treasury however he saw fit, without having to consult anyone. The religious authorities of al-Azhar preached obedience to his will. There were no elections. Villages were governed by shaikhs whom he could replace on a whim, and provinces were overseen by elites that depended on him and him alone for patronage. The army was his to command, and its generals served at his pleasure.

And at no point did Said contemplate political reforms that would have diminished his authority. On the contrary, his policies were designed to centralize authority and enhance his power, in order to make it simpler for him to govern and to raise revenue. The country had been divided into administrative districts by his father, in imitation of the Napoleonic reforms that reorganized France. Said went a step further and extended these reforms to the bureaucracy. He was the first Egyptian ruler to organize a government with civil servants who received published salaries. As radical as that was, salaries were still open to negotiation based on favor and bribes, and the civil service did not have any authority separate from the will of the viceroy. Above all, Said was skeptical of the populist spirit of the 1848 revolutions. He believed in educating the ruling class so that they could rule better, but he saw no reason for the peasants to learn to read. "Why open the people's eyes?" he explained. "That only makes them harder to rule."[1]

Said's ideology manifested itself in his physical surroundings. His court was an eclectic mix of Eastern ostentation and Western conveniences. The Ras al-Tin Palace combined cedar beams from Lebanon with gold-gilt work that echoed Versailles. Persian rugs designed to cushion divans surrounded sofas and chairs made in Europe. The evening entertainment consisted of singing girls and musicians who played reed fifes and the *darabuka* fish-skinned drum, while the guests washed their hands in European finger-bowls. The army was dressed in uniforms adopted from the Prussians and Austrians; the members of the court wore loose, richly embroidered robes and shawls, sometimes topped by a red Turkish tarboosh, or by a white lace skullcap. Said himself often sported a fiery ruby ring on his finger, along with a tarboosh encrusted with jewels. He loved to entertain, and much of the business of his court was conducted not in formal sessions but during long evenings of song, dance, and food.

And Europe responded, both socially and economically. The accession of Said was a klaxon for European entrepreneurs who had been stymied during the years of Abbas. Projects that had been shelved were revisited. Under Abbas, foreign consuls were rarely invited to the court, and their social lives dimmed. Thousands of Europeans who had flocked to Alexandria during Muhammad Ali's reign in search of new horizons went elsewhere to seek their fortunes. But in the fall of 1854, that stream reversed course, and the expatriates returned in the hope of making deals.

The consuls were the key middlemen. As diplomats, they guarded their country's interests and made sure that their countrymen were treated fairly. But they were also free agents who enriched themselves by selling access. The French consul might set up a meeting between a French railroad consortium and a member of the royal household who owned land along the Nile. The consul would take a fee from each party, most likely in the form of stock in the joint venture. He would then have a vested interest in seeing the project completed, but no costs if it wasn't. Each country had consuls, and the ultimate goal was getting the viceroy to grant a concession. Even though Said was generous in his awards, there were far more people jockeying for an audience and for influence than there were available projects. Competition was fierce, but the potential profits were worth the trouble.[2]

Said was accommodating. He was perceived as an ingenuous manchild who could be easily tempted by lavish descriptions of new technologies. In front of one of his palaces was a broad open space where he liked to conduct troop maneuvers. During the summer, it was impossible to use because of the excessive dust. A European merchant convinced Said that the problem could be solved by covering the courtyard with iron. The expense was prohibitive, but Said agreed to it. The space was paved, which eliminated the dust issue. The metal became so hot in the summer sun, however, that it burned the feet of the soldiers through their shoes, and the iron eventually had to be removed. Because of unwise decisions such as these, Said gained a reputation as an easy mark for a clever entrepreneur with a scheme.

Some of that reputation was deserved. He gave out concessions at a far greater rate than either Abbas or Muhammad Ali, and tended not to assess the financial risks beforehand. But Said also was plagued by avaricious European consuls and their business allies, who took advantage of the power disparity between their governments and Said. Rather than treating the concessions as glorified permission slips, consuls and businessmen argued that they were actually contracts between the Egyptian government and private parties. That meant that if the project did not come to fruition the Egyptian government could be held partly liable and required to indemnify the concessionaire.[3]

Though he did not agree with this interpretation, Said lacked the power to reject it. Beginning with Napoleon, Europeans had extended their extraterritorial rights, and had managed to place themselves under the legal umbrella of their consular representatives rather than the Egyptian government. These arrangements, known as capitulations, had begun centuries earlier as agreements between the Ottoman Empire and the states of Europe. Initially, they had been negotiated from a position of strength by the Ottomans, but as the balance of power shifted, European governments were able to demand more privileges. By the time of Said, the situation had reached an absurd level. Europeans could rarely be tried in Egyptian courts, no matter how local the crime. Instead, they were turned over to their consuls, who could choose to prosecute or not.[4]

In business deals, the capitulations often gave the Europeans carte blanche. Few consuls had an interest in granting Egyptian claims against European entrepreneurs. If Said wanted to deny a European claim, he had to appeal to third-party arbiters, and often these were European aristocrats, more likely to sympathize with the way the consuls interpreted contracts than with the viceroy of Egypt.

Said was aware of the risks. He didn't trust the consuls, and once referred to them as "wolves." But he took the chance of incurring excessive costs because he saw no other viable path. He was in the same quandary as dozens of other rulers of non-European countries: he needed European capital in order to change his country enough to resist European influence. It was a problem without a good solution, and one that bedeviled Said's successors in the twentieth century as much as it compromised Said himself in the nineteenth. The peoples of Egypt, of the Middle East, and of much of the world outside of the United States and Europe have been struggling to catch up for more than a century. It was frustrating for Said when the game began. A century and a half later, that frustration gave way to bitterness, and to hatred.

In 1854, however, Said looked to a brighter future. He was young; he had a zest for life; and he was the ruler of an ancient country blessed with fertile soil. He also could draw on a reservoir of good will in

Europe. Compared with Abbas, any successor might have been greeted warmly, but Said had made many friends in Europe and among the European community in Egypt and was hailed as the agent of a new and better future.

Ferdinand de Lesseps was therefore only one of many who saw the elevation of Said as a wonderful opportunity. Hearing that Said had assumed power, Lesseps wrote to him immediately. Said had not yet developed that reputation as an easy mark, but Lesseps had known the prince on and off for twenty years, and must have recognized that Said could be persuaded to endorse an idea that had been rebuffed by his predecessors.

While running his mother-in-law's estate in Berry, Lesseps had been searching for a grand project. Though the Suez Canal was only one of the many possibilities, there is no clue as to what the others might have been. Lesseps is the sole source of information about his activities in the years following his disgrace, and he left no record of what other plans he might have considered. He had been following the progress of the Study Group since its inception, and he was in occasional contact with several of its members. He also carried on a correspondence with the Dutch consul general in Egypt, M.S.W. Ruyssenaers, an old friend who knew everything and everyone in Alexandria and Cairo. Since at least 1852, Lesseps had been analyzing the surveys and blueprints assembled by Enfantin's group, and he had been one of many who lamented that Abbas had closed his mind to European influence. In April 1853, he wrote to Abbas in a futile attempt to convince him to undertake the construction of the Suez Canal. In arguments that would later prove persuasive to Said, Lesseps spoke of the grandeur of the project and its financial benefits, but Abbas was unmoved.[5]

It seems that Lesseps made these overtures on his own. Without the earlier work of the Study Group, he would have had a hard time acquiring such a keen grasp of the relevant details, yet the exact nature of his relationship with Enfantin and the Study Group is sketchy. According to his competitors, Lesseps stole both the idea for the canal and detailed surveys that had been compiled by the Study Group. According to his partisans, the general idea of a canal was common knowledge, and Lesseps simply took it upon himself to implement it. Had Abbas remained in power, this chicken-and-egg question would have been moot. But events soon made it central.

From the moment he heard about the death of Abbas, Lesseps

planned to go to Egypt and present the canal idea to Said. He wrote to Said in late July. He received a response a month later, warmly inviting him to visit, and he left France in late October 1854. What he did then is open to debate. He may or may not have stopped in Lyon on his way to his ship in Marseille, and he may or may not have visited the offices of François Arlès-Dufour, one of Enfantin's closest aides. And while in Lyon, he may or may not have been given documents that he then used as source material for selling the project to Said.[6]

There is no way to know for certain. Lesseps always maintained that he drew up plans on his own, but, given the later acrimony between him and Enfantin, that is precisely the position he would have been expected to take. In a similar vein, Enfantin's self-interest would dictate the alternate story. However Lesseps obtained the preliminary plans, by the time he arrived in Alexandria in November he was armed with information on the possible canal and options for various routes. The Study Group had been unable to arrive at a consensus over which path was best, but they had at least narrowed the choice to two options: either a direct route from the Mediterranean, cutting through the isthmus to the port of Suez, or an indirect route using the Nile and then fashioning a ship canal from the outskirts of Cairo to Suez. Lesseps was certain that only the direct route made sense, and he refused to consider any alternatives. From his perspective, the indirect route presented political and financial obstacles that would doom the project to failure.

Disembarking in Alexandria, he was met by Ruyssenaers and a representative of the viceroy, who escorted him to a palatial villa near the Mahmoudiah Canal, several miles away. Though he was a private French citizen with a comfortable but by no means substantial income, he was treated as visiting royalty. A phalanx of servants lined up to greet him at the main entrance, all Turks and Arabs save for a Greek valet. The palace had recently been used by one of Said's wives, who had borne him a son. It was outfitted with every convenience, and every luxury: silk bedding, brocaded upholstery, gold-embroidered towels, marble dressing rooms, and a set of servants whose sole responsibility was to prepare his coffee and supply him with water pipes.[7]

Informed that the viceroy was ready to receive him, Lesseps dressed in his most formal attire, replete with the medals and orders of merit he had been awarded during his years in the diplomatic service. He wrote to Madame Delamalle that his reason for dressing that way was to show

Said "respectful deference," but the actual reasons were probably more complex. As a private citizen, he had no actual standing in Egypt, and though he had warm memories of his friendship with Said, he didn't want to rely purely on good will. By festooning himself in the regalia of a high official, he may have wanted to give the impression that he had more influence in France and in Europe than he actually did.

In addition to his history with Said, Lesseps had one other asset. His cousin Eugénie de Montijo had recently become Empress Eugénie, and though his degree of closeness with her was a matter of gossip and speculation, Lesseps's exact position in the international hierarchy must have been difficult for Said and his court to gauge. Just as Lesseps thought it best to give the impression of power that he didn't truly possess, Said thought it best to treat his old friend as someone who did.

The two men met for the first time in fifteen years in the formal audience chamber of the Ras al-Tin Palace, and according to Lesseps, it was as if only a short spell had passed since they had last seen each other. Lesseps, who had developed the silken tongue of a diplomat after long years in that profession, showered the new potentate with compliments. Said, in turn, reminisced about his youth and thanked Ferdinand for supporting him even at the risk of incurring Muhammad Ali's displeasure. He then invited Lesseps to come with him on a trip across the desert to Cairo. Military parades were planned, and Said hoped that Lesseps would be his honored guest. Ferdinand, who was looking for such a chance to be alone with Said, said yes.

Knowing of Said's passion for shooting and hunting, he had brought a pair of revolvers from France as a gift. He saw the viceroy several times before the trip began, once as a guest during the formal audience hours, and once in Said's private apartments for a desultory few hours spent reclining on divans overlooking the gardens of the palace while languidly smoking and drinking coffee, the caffeine doing battle with the light-headed stupor that washed over them as they inhaled the pipes. They talked about many things, but Lesseps did not raise the matter of the canal.

Furnished with an Arabian horse by Said, Lesseps joined the royal party on its journey south on November 13. Winter was setting in; though the days would have been temperate, the nights in the desert were cool. After crossing the lake separating Alexandria from the mainland, they were soon surrounded by the sands of the Libyan Desert, where Napoleon's troops had slogged in parched determination half a century before. Lesseps rode on his own for a time, and then arrived at the camp that evening. Tents had been pitched in neat rows, and the imperial compound mimicked the opulence of the palace they had just left. Lesseps's tent was furnished with an iron bed covered in silk. He was then taken to a larger tent where the viceroy and his court were gathered. Food was served on Sèvres china, and ice water was provided. The army, meanwhile, had set up artillery pieces for demonstrations, and in between the booming of the cannons, the imperial party received news of the siege of Sebastopol. England and France in alliance with the Ottomans were fighting a war against Russia for murky reasons on the Crimean Peninsula, but it was the only war going on, and news of it was prime entertainment. The conversation rambled for hours, and at one point Said talked excitedly about his desire to do something dramatic to mark the beginning of his reign.

Two days later, just before dawn, Lesseps woke up, wrapped himself in a red dressing gown, and peeled back the flap of his tent. He gazed on the desert. "A few rays began to light up the horizon," he wrote. "To my right, there is the east in all of its limpidity; to my left, the west is dark and cloudy. All of a sudden, I saw a rainbow, with all of its vivid colors, spanning east and west. I swear, I felt my heart beat faster, and I had to stop my imagination from seeing this as a sign of the alliance spoken of in the Scriptures, of the true union between the West and the East, and as a message that today would see the success of my project."[8] Harbinger or not, Lesseps took the rainbow as a cue to broach the subject of the Suez Canal.

Toward the end of the day, he mounted his horse and galloped to Said's tent. The viceroy was alone, save for his servants, and he invited Lesseps to sit for a while. Ferdinand had rehearsed what he planned to say, and he decided that, rather than going into details, he would present Said with a vision. He spun a tale of the ancient canal, of the pharaohs and Alexander, of the Arab conquerors and Napoleon. He told Said that the time had come to fulfill their legacy by cutting a direct canal across the Isthmus of Suez.

He then presented the reasons: A canal would strengthen the Ottoman Empire and its sultan, who was still Said's sovereign. A canal in Egypt would guarantee Egypt's independence, just as the Dardenelles and the Bosporus guaranteed the autonomy of the Ottomans; all European powers would covet access to the canal, just as they did to the Dardenelles straits that connected the Black Sea to the Mediterranean. Rather than allow any one power to control these vital waterways, the states of Europe would guarantee the security of a neutral party, and that meant that Said would never have to fear a European invasion. The canal would show that Egypt "still has the capacity to be a potent force in world affairs, and is still capable of adding a brilliant page to the history of world civilization." He promised that, though the technical obstacles were formidable, they weren't insurmountable, as Said's own chief of public works, Linant Bey, had demonstrated. The real challenges were political and economic, but Lesseps assured the viceroy that, with effort, both could be met.

This was only a prelude to the real lure. A canal, continued Lesseps, would set Said apart from other rulers. It would transform him from the governor of an Ottoman province into a potentate admired throughout the world and immortalized as a man who dared to do what others had said was impossible. "The names of the Egyptian sovereigns who erected the Pyramids, those useless monuments of human pride, will be ignored. The name of the prince who will have opened the grand canal through Suez will be blessed century after century for posterity."[9] The canal, Lesseps continued, would secure the passage of pilgrims to the holy sites of Mecca and Medina and make the ruler of Egypt a protector of the faithful. The canal would connect Europe to the lands of India, China, Japan, and Australia, and place Egypt at the center of world trade. It would enrich Egypt and bring it more revenue than cash crops ever could.

The combination of glory and money was hard to resist. It had, after all, enticed Lesseps, and as good a salesman as he was, he had the advantage of selling a vision that he himself fully believed. At the end of the presentation, Said turned after some thought and said, "I am convinced. I accept your plan; for the rest of the trip, we will figure out the actual means of implementing it. The matter is settled, you have my word."

With that late-afternoon conversation, the Suez Canal ceased to be only a dream. After half a century of stops and starts, the impasse was finally broken by two people talking in the desert. Of course, these two were actually in a position to do something other than talk, or at least one of them was. But that was all that was required. The sovereign of

Egypt gave his personal word to a private citizen of France, and in that moment, everything changed.

Of course, there was still the matter of actually building the canal. It was easy enough to spin webs of fancy in the crisp air of the Libyan Desert. If every brilliant idea hatched over drinks came to fruition, the world would have been a perfect place long ago. Said had given his word that night in November, but no one could have held him to it. The dream could have faded the next morning, never to be revived. Said alone would only have pursued the plan so far. He had neither the money nor the expertise to see it to completion. No matter how keen he was on the notion, it required European capital and European engineering, backed up by Egyptian workers. Though we have only Lesseps's account of their meeting, Said must have been enthusiastic. At first glance, he could see only the upsides and none of the pitfalls. The project would be organized by others and financed by others, while the glory and the influence would go to Egypt and its ruler.

Said's father would have scoffed at that thought. Muhammad Ali understood that effortless glory was an oxymoron, and he had refused to sanction the canal because he recognized the risks. Said, more trusting and less experienced, focused on the rewards, as Lesseps knew he would. That became a pattern for the viceroy, and his audience chamber soon teemed with opportunists hoping to obtain promises from Said that they could then record on paper and enforce as contracts. Lesseps himself went back to his tent that evening and formalized his notes into a memorandum that he gave to Said the next day. A tendency to think big without pondering the consequences has cost Said in the docket of history. He has been charged with naïve profligacy and found guilty. It is impossible to absolve him fully, but many of his decisions looked sound at the time. His treasury was stable; his foreign debts were nonexistent; and his sense of how Egypt needed to change was largely accurate.

Lesseps, on the other hand, seemed to be making things up as he went along. He knew enough about Said's character and about the vanity of rulers to strike the right notes. In fact, one of Lesseps's greatest assets in the coming years was his ability to persuade. His language was florid, but that was the style of the day. He spoke in hyperbole, but he lived during a time when words such as "destiny," "glory," "progress," and "civilization" were commonplace. What set him apart was a fierce determination to do something meaningful, combined with astonishing energy, good connections, and a spellbinding ability to convince people that he could change the world for the better if only they would help him.

From that day in November 1854, it would take fifteen years of ceaseless effort to create the Suez Canal. Had Lesseps flagged for one moment, the project could have collapsed. Not until the early 1860s had enough actual physical work been done on the canal to ensure that it would be finished by someone, even if Lesseps's venture disintegrated. Said's support was an imperative first step, but it was only the beginning of a long, uncertain road that time and again looked as if it would end in failure.

That day also marked the transformation of the Suez Canal from an idea shared by many to a project led by one. From 1854 onward, it became Ferdinand de Lesseps's canal, and he had no intention of sharing. He was determined to occupy center stage, and he was willing to take the risk of full blame if that meant the possibility of full credit. But the other contenders did not simply move aside when they learned of what had transpired in the desert. Some of them were furious that Lesseps had broken the rules of the game and gone straight to the viceroy rather than consulting with various interest groups in Europe and Constantinople first. And at least one of them was as determined as Lesseps to wear the laurels.

CHAPTER SEVEN

# WHOSE CANAL?

H AVING OBTAINED THE viceroy's verbal approval, Lesseps spent
another week traveling with the royal party to Cairo. The days
were filled with military maneuvers designed more to entertain
the royals than to drill the army. The regiments were arrayed in full
dress. They marched, and shot, and arranged themselves in columns,
but there was at least as much precision in the kitchen tents. Lesseps
marveled over the dozens of cooks and their elaborately choreographed
meals.

During the trip, Said received the shaikhs of the Bedouin tribes that
controlled the desert lands to the west of the Nile and north of Cairo.
They galloped around his tent and fired carbines in the air to display
their strength, but it was all posture. They were vassals of the viceroy,
and they paid him tribute. In return, he made a show of respect and
willingly played his part in these elaborate rituals.

The last days of the trip were completed aboard Said's Nile yacht,
which he had inherited from Abbas, and which had been custom-made
of citron wood and oak in England for a rumored cost of one hundred
thousand pounds. Inside, there was a dining room forty feet long, and
divans lined with gold. Said defensively claimed that he would never
have spent so much on such a frivolity, but, seeing as how he had
inherited it, he thought it best to put it to good use. Lesseps wisely
demurred, and took advantage of the immense guest-room he was
given, complete with divans upholstered with silk and a bathroom of
white marble.

By late November, the party arrived in Cairo, and once there,
Lesseps made sure to solidify Said's agreement in writing. He had
already talked up the Suez idea with the viceroy's staff and inner circle,

including Ahmad Pasha, Said's nephew and heir, and Mustafa Pasha, Ahmad's brother. The moment he settled in Cairo, Ferdinand wrote to each of the European consuls to inform them of what had transpired on the trip. On November 27, he wrote to Frederick Bruce, the British consul general (and son of Lord Elgin, of the Elgin Marbles), to announce that Said had authorized him to form a company for the purpose of piercing the Isthmus of Suez. He said that there would be a formal concession granting the company lands for income and guaranteeing a supply of labor in the form of the traditional corvée. Lesseps promised that the canal would benefit not just the company but international trade, and therefore England most of all. But, recognizing the legacy of tension between France and England, Lesseps asked Bruce "to consider as a heresy the belief that this enterprise—which is destined to cut in half the distance between the Orient and the Occident—will not benefit England—lord of Gibraltar, of Malta, and of Aden, and of important bases on the east coast of Africa, of India, of Singapore, and Australia." He concluded by saying that anyone "preoccupied with questions of civilization and progress cannot look at a map and not be seized with a powerful desire to make disappear the only obstacle interfering with the flow of the commerce of the world."[1]

The point about the map was telling. For years, Lesseps sat in his farmhouse in the French countryside and studied plans. He laid out maps on tables, desktops, and floors in order to plot various paths. Until he returned to Egypt in the fall of 1854, the entire scheme had been no more than a map game, and he had not yet gone to the isthmus to see the potential route for himself. It was still a drawing to him, and that is how he described it to Bruce and to others.

Within days, Lesseps had his much-coveted piece of paper, with the grand label of "concession." Granted by "His Highness Muhammad Said, Viceroy of Egypt," it gave "our friend M. Ferdinand de Lesseps . . . the exclusive power of constituting and directing a universal Company for cutting through the isthmus of Suez and establishing a Canal between the two Seas, with power to undertake or cause to be undertaken, all works and constructions." Lesseps was made the director of this soon-to-be-constituted company, and the length of the grant was set at ninety-nine years. The company was to be responsible for all costs of the project, but "the Egyptian government is to give up those portions of the public property now uncultivated which would be watered and cultivated at the expense of . . . the said Company. The Company will enjoy possession of the said property for the term of ten years from the day of the opening of the canal." For the remaining eighty-nine years of the concession, the company would be required to pay a tithe to the Egyptian government.

The other major provisions of the concession included an agreement that the Egyptian government would receive 15 percent of the company's annual net profits, with 10 percent for the company's founders and the remaining 75 percent going to the company's shareholders. Linant Bey was designated the official engineer; he would draw up plans for the canal's route and determine which lands would be granted to the company. Working closely with him was another transplanted Frenchman, Mougel Bey. A hydraulic engineer by training, Mougel had come to Egypt in the late 1830s to oversee construction of the Nile dams, and, graced with a generous salary from the Egyptian government, he had created a comfortable life for himself. Though he had worked closely with Linant for years, the two men had an uneasy relationship, perhaps because they frequently competed for the same projects.

The concession did not stipulate whether the canal would be a direct line through the isthmus, or whether it would follow the indirect route via the Nile. Linant had always preferred the direct line, but at this juncture the question was left open to further study. Either way, the company was given the right to import any necessary equipment, mine any quarries, and bring in materials free from taxation by the Egyptian government.[2]

Even at the time, the concession struck many as a blank check for Lesseps and an act of misguided generosity on the part of Said. Certainly, the agreement had been drawn up quickly, and mostly by Lesseps himself. The concession of land would prove to be the most controversial aspect, though granting an unformed company unirrigated land must have seemed to Said innocuous enough. Only after a freshwater passage had been dug from the Nile to Lake Timsah at the canal's midsection would those dry sands become some of the most fertile land in Egypt (and indeed the world) and thereby raise the value of those acres immensely. But, much as the United States government granted railroad companies extensive lands on either side of the proposed tracks to be laid, the concession of land to the Suez company was not atypical for the era. And it was economically necessary. Such a vast undertaking could only be partly underwritten by wealthy individuals and shareholders. Other income would have to be found before the project was completed, and in the mid-nineteenth century, it was uncommon for speculative ventures to be funded by bank loans. Banks gave loans to governments and to merchants engaged in trade, but not usually to entrepreneurs.

That would soon change. One of the innovations in France in the early 1850s was the creation of a new type of finance company, known as the Crédit Mobilier. In the past, there had been joint-venture companies meant to fund speculative projects, but the Crédit Mobilier was one of the first to resemble the modern investment bank. It allowed for the financing of projects that were long on vision and short on collateral, and its investments underwrote a railroad boom in France and the massive renovation of Paris under the direction of Baron Haussmann. As fate would have it, the Crédit Mobilier was a product of the Saint-Simonians, just as the idea for the canal was.[3]

The brothers Émile and Isaac Pereire broke with Enfantin in the early 1830s, along with many of the initial followers of Henri de Saint-Simon. But they did not abandon the doctrines of Saint-Simon, one of which held that banking and credit were integral components of industrial progress. In that spirit, they set up the Crédit Mobilier as a vehicle both to enrich its shareholders and to fund grand public works.

While the Pereires plunged into the world of finance, another wing of the Saint-Simonians continued to follow the now wealthy but still-eccentric Enfantin. Though he had mellowed somewhat with age, he was no less given to flights of rhetoric that placed him at the center of the cosmos, and in dealing with the noninitiated, his loose confederation made sure that others spoke for him. One of these spokesmen was Arlès-Dufour, born François-Barthélemy Arlès. As a young man, he settled in Lyon, then a flourishing industrial center and one of the major cities of France. He lived there the rest of his long life. Physically, he was pleasantly nondescript, and in his various portraits he appears prosperous, kind, and slightly sad.[4]

Rising to become head of the Lyon Chamber of Commerce, Arlès-Dufour was involved in countless projects until his death in 1872. For five decades, he was a quiet but staunch follower of Enfantin. But as a moderate soul, he was able to bridge the often considerable distance between the Father and most others. After all, the majority of French and European businessmen agreed that progress, industry, and profit were wonderful, even if they were skeptical about tales of the male and female life-force.

Under Arlès-Dufour's initiative, the Lyon Chamber of Commerce made a contribution of five thousand francs to Enfantin's Study Group, and it was Arlès-Dufour who became the main correspondent of Ferdinand de Lesseps in the years between his disgrace in Rome and his voyage to Egypt. Ferdinand seems to have considered the Lyon businessman a friend, at least judging from the warm enthusiasm of his correspondence with Arlès-Dufour in the flush of success at the end of 1854 and into 1855. But while Lesseps was writing to Arlès-Dufour as a trusted ally, Arlès-Dufour was also carrying on a separate correspondence with Enfantin, and the latter did not share the perception that Lesseps had done a great thing for a project that had long lain dormant.

The same day that the canal concession was formally awarded, Lesseps sent a letter that began, "My dear friend Arlès-Dufour, Rejoice—you and our friends—Rejoice! I have succeeded today in obtaining everything that I had hoped for." He told Arlès-Dufour that the document had been signed, and that he was already onto the next stage of plans. Though he didn't know for certain how much it would cost, he said that the company would be capitalized at two hundred million francs. That figure was predicated on the assumption of a direct route; in theory, all options for possible routes were supposed to be considered open, but Lesseps told Arlès-Dufour that the only thing he was interested in was the direct path. To that end, he had spent many hours with both Linant and Mougel going over plans and preparing for a trip with them to the isthmus in a matter of weeks. The expedition would be paid for by the viceroy in his capacity as one of the founding shareholders.

Lesseps asked one favor. He wanted Arlès-Dufour to begin contacting the major banking houses of France and Europe, including Benoît Fould and the Rothschilds, and to make overtures to the "great capitalists of England." For this, Arlès-Dufour would be well rewarded. Lesseps had listed him as a founder, and that was more than just a paper honor. By the act of concession, 10 percent of the canal's profits would be set aside for the founders, who included Said and multiple members of his family, Lesseps and multiple members of his, Linant, Mougel, the Dutch Consul General Ruyssenaers, the French Consul General M. Sabatier, the brothers Talabot, Benoît Fould, the English free-trade advocate Richard Cobden, members of the chambers of commerce of Lyon and Venice, and Enfantin himself.[5]

Many of these people were included before they had actually been approached and asked if they wanted to be. Lesseps preferred to err on the side of listing too many rather than too few. He understood that even someone such as Baron James Rothschild would be tempted, flattered, and seduced by being selected as a founding member of a great venture, and thereby be more likely to support the project. For that reason, the list contained possible adversaries and potential patrons along with established friends and supporters.

Remarkably, though it would take another five years of diplomatic wrangling before work on the canal actually began, the rudimentary blueprints drawn up by Lesseps, Linant, and Mougel in these first weeks formed the eventual template. Dozens of other engineers would survey the isthmus in later years in order to validate or debunk the feasibility of the route charted in these first weeks, but the final project looked much as Lesseps thought it would at the end of 1854.

Lesseps had to function at several levels. He faced an assortment of hurdles, and they each required careful navigation. At the diplomatic level, there was the question of which governments would support the canal and which would not. The concession may have called for the canal to be built by a private company, but a project of this nature could not avoid becoming political. The states of Europe, and England and France especially, were actively competing in the world outside of Europe, and anything that could materially alter the balance between them would become a contentious issue. The canal promised to change the dynamics of international trade and to redirect the flow of goods in ways that would cause all concerned countries to turn their attention to Egypt. At the outset, Lesseps anticipated that France would support him, and that England would not. And, as much as he needed the good will of the Emperor Louis-Napoleon, the acquiescence of Britain was even more vital.

The Ottoman Empire was also a central player. Legally, Said ruled Egypt at the whim of the sultan, and he had no authority to "alienate" Egyptian land without the sultan's permission. By granting the Canal Company land rights, Said ventured into questionable territory, and the first act of concession was silent about the sultan. Lesseps no less than Said recognized that the project could not continue indefinitely without consulting the Ottoman government in Constantinople, and Lesseps planned to head there as soon as he had made a preliminary survey of the isthmus. Whatever he believed about the legal relationship between the viceroy and the sultan, Lesseps argued that precedent gave Said the right to undertake the canal project without prior approval from Constantinople. Gaining such approval would be useful, and appealing to the sultan was at the very least good manners and good politics. But Lesseps refused to endorse the principle that Said required the permission of the sultan, and doing so would have transferred the political question out of Egypt and into a more complicated arena, where Lesseps would not enjoy the same advantages.[6]

Then there were the technical obstacles, which had to be resolved even before the equally challenging financial hurdles could be approached. Ever since the renewed flurry of interest in the canal in the 1840s, opinion among the engineers of Europe had been divided about the feasibility of the idea. No canal of this size and scope between two seas had ever been constructed. Knowledge about the region was scant. How would the sands be kept at bay? What would prevent the head of the canal from silting up? Were the sea levels really the same? Would the canal overflow its banks? And what about the currents of the Red Sea? But the most pressing question was whether the route should cut directly through the isthmus or use the indirect path via the Nile. That was both a technical debate and a political one. Those favoring an indirect route tended to represent the merchant community in Alexandria, which stood to lose greatly if the canal bypassed them. Those favoring the direct path saw it as cheaper, simpler, and more efficient. A direct route would also offer better competition to the railroad then being built from Alexandria to Cairo under English auspices with Egyptian government funding. Said himself had an additional reason for preferring the direct route. Unlike Muhammad Ali, who wanted the canal where he could keep an eye on it from Cairo, Said wanted the channel to be distant from the power centers of Egypt. In theory, that would prevent the meddling states of Europe from interfering in domestic Egyptian affairs. In reality, it did not.

Lesseps had his own reasons for championing a direct route. Others had already staked their reputations on the indirect path, and if that was adopted, it would be almost impossible for Lesseps to claim sole credit for the canal. He did not admit to this motive; he may not even have been aware of it. But, self-consciously or not, he sought more than full control; he wanted absolute credit, and he made herculean efforts to appropriate the legacy for himself alone.

There were several other claimants, but the most organized and formidable were the remnants of the Study Group assembled around Enfantin. By writing regularly to Arlès-Dufour, Lesseps both kept the Study Group appraised and signaled to Enfantin that the canal was now his. If any of the former members of the Study Group wished to participate in the project, they would have to do so on Lesseps's terms. Over the next months, there were heated debates in the capitals of Europe over which plan for the canal was the most prudent. Learned men and assorted entrepreneurs published opinion pieces, gave speeches at meetings, and argued in drawing rooms and salons about the technical pros and cons of the various options. But the most heated debate was over control, and the technical debate was merely a proxy for a contest for fame.

In the end, reputation is what mattered for Lesseps, far more than the money. He wanted to perpetuate his name and have it associated with progress and civilization. It is true that he enjoyed the finer things in life, and lovingly described the opulence of his surroundings wherever he went, but he wasn't in this game because he craved lucre. He was driven, fiercely and inexhaustibly, by a passion to be known and to have the world say "Ferdinand de Lesseps" with that mixture of awe and admiration that is reserved for a very few.

Over Christmas 1854, Lesseps left for the Red Sea port of Suez with the French Consul Sabatier, Madame Sabatier, Linant and Mougel, an escort of armed horsemen, servants on camels, an assortment of soon-to-be-eaten sheep, goats, and pigeons, and dozens of barrels of Nile water. Lesseps then spent three weeks traveling north toward the Mediterranean. He found the town of Suez desolate and its three thousand "miserable" inhabitants eking out a bare existence. Near the town were the outlines of the canal bed described by Herodotus and other travelers, and it was there that Lesseps began the survey.

As this party made its way along the proposed route, Linant offered his advice about canal building and Mougel talked about hydraulic engineering. After a few days assessing the quality of the stone in the mountains of Attaka, Lesseps and the engineers parted with Sabatier, mounted their camels, and began to head toward the Bitter Lakes, which had long since dried up and were now large, empty, salinated

depressions filled with eerie crystalline formations. The region was
uninhabited, though Lesseps did note telltale signs of desert gazelle,
antelope, and hyenas. Skirting the jagged lake-beds, they crossed the
Serapeum plateau and into the region of Lake Timsah.

During the trip, Lesseps began to read the Biblical story of Moses,
and he was swept up in the headiness of following in the footsteps of
the Hebrews. Europeans who explored the lands of the Near East had
tried to match the stories of the Bible with physical places, but the
question of where Moses and the fleeing Jews of Egypt had crossed
into the Promised Land was unresolved. The Bible says that the Red
Sea parted, but scholars doubted that the Hebrews had gone so far
south. Lake Timsah, in the middle of the Isthmus of Suez, or the Bitter
Lakes were more likely alternatives, since they were smaller and
offered a more direct route into Israel. And because they had been
filled with salt water from the Red Sea, ancient writers might have considered them part of it.

As Lesseps read his Bible he wrote in his journal about walking
where Joseph had been enslaved and where Moses had wandered. He
compared the place-names in the Bible with the current names given
those landmarks by the local Bedouins, and added his annotations to
Biblical passages. For instance, in the Bible it says, "The children of
Israel journeyed from Rameses to Succoth," to which Lesseps added:
"Succoth in Hebrew means Tents. This place is now called by the Arabs
either Oum-Riam (the Mother of Tents) or Makfar (the Hollowed-Out
Place where the old canal passed)." But though he wrote these descriptions, he was guided by Linant, who had traveled through the region
numerous times and had spent years trying to correlate its geography to
the stories of the Bible. Linant was convinced the area around the Bitter Lakes and Timsah formed the Biblical land of Goshen, and that the
low areas near Timsah were called the Valley of the Reeds; this may in
turn have been why translations of the Bible incorrectly referred to the
"Red Sea" instead of to the "Reed Sea," which would have meant the
reedy Lake Timsah.[7]

The Bible aside, Lesseps was pleased that Lake Timsah was large
enough to act as a midway port for the canal, as well as a logical spot
for the intersection of the freshwater canal from the Nile. He took
strength from the thought that the land of Goshen, described in the
book of Genesis as a region of fertility and abundance, would bloom
again. Lesseps was not particularly religious, but, like many of his

upbringing and class, he saw the Bible as the cornerstone of Western civilization and the ancient Hebrews as vital actors in God's plan to raise the West to the pinnacle of world culture.

From Timsah, the group continued north to Lake Manzala, through oases of date palms and past marshes filled with swans and pelicans. It took only two days to go from Timsah past the plateau of El-Guisr and then to the ruins of Pelusium on the coast. On the way back to Cairo, Lesseps was struck by the legacy that he was inheriting, not just from the ancient world, but from his father. He wrote with unusual humility in his journal that he was now completing a saga that had been begun by his father and Muhammad Ali, and which now seemed to be reaching its glorious culmination at the hands of both of their sons.

The expedition returned to Cairo in mid-January 1855, and Lesseps began another intensive round of correspondence. News of the canal had spread quickly, and interested parties in Europe had begun to take their stands. Lesseps understood that he could not dawdle, and that adversaries and competitors would take advantage of any delays to wrest control or elevate their schemes over his. In Cairo, he instructed Linant and Mougel to explore the issues raised by the brief survey in the isthmus. What work needed to be done at the port of Suez? How should the canal use the dry basin of the Bitter Lakes? What should be done with Lake Timsah? What exact path should the canal follow from Timsah to Lake Manzala? How could the marshy area near Pelusium be transformed into a suitable port? How long would the jetty have to be from Lake Manzala into the Mediterranean? How would the tides affect the work? What about the dunes and sands in the desert regions? And, finally, how much would it cost? He told them to spare no expense in answering these queries, and asked only that if they differed about the answers they would keep such disagreements secret so as not to give the enemies of the canal any ammunition.[8]

Lesseps then left for Constantinople to explain the project to the government of the sultan. He also began to assemble material to refute the growing chorus of arguments against the canal in general and against a direct canal, many of which were emanating from England, but one of which was coming from an unexpected direction.

He continued the regular exchange of letters with Arlès-Dufour. At first, he seems to have assumed that Arlès-Dufour, Enfantin, and the Study Group would embrace him. After all, he had won the concession that the Study Group had tried to obtain years before, and which

Enfantin had yearned for since the 1830s. But it soon became clear that Lesseps's success had engendered envy and rage, rather than warmth and enthusiasm. Having been apprised of Lesseps's plans by Arlès-Dufour, a very grumpy Enfantin used his formidable connections to wangle an interview with Napoleon III.

As a young man, Louis-Napoleon had dabbled in the ideas of Saint-Simon, and he was rumored to be a closet Saint-Simonian. Meeting with the emperor, Enfantin was annoyed but deferential. He wanted Napoleon to support the Study Group and not Lesseps. He alluded to the familial connection between Lesseps and the new Empress Eugénie de Montijo, and he implied that the emperor ought not be guided by that. Enfantin believed that the Study Group had a prior and valid claim to the project, and that, although Lesseps was welcome to participate, he was not entitled to call the enterprise his own.[9]

During the previous weeks, Enfantin had written to Lesseps through Arlès-Dufour. He was not pleased. He objected to the use of Pelusium as a port and recalled that the Study Group had rejected Pelusium for good reasons. It was too shallow, and though the mouth of the Nile had in past centuries moved west, the alluvial sands would still be a chronic problem. In addition, there were no quarries in the vicinity, and the cost of transporting the massive quantities of stone needed for the extensive jetty would make the overall cost of the project prohibitive. Enfantin referred to the studies made by Paulin Talabot in 1847 for proof that the design now championed by Lesseps was seriously flawed, and he argued instead for a version of the indirect route that Talabot had recently gone on record supporting.

But Enfantin's real gripe was that Lesseps had taken charge of an endeavor that wasn't his to lead. He felt that the Study Group had a prior claim, though he acknowledged that the opposition of the English member of the group, Robert Stephenson, was a potential liability. Enfantin talked to Lesseps as if to an old friend with whom he was having a misunderstanding, and he asked Ferdinand to respect the prior rights of the Study Group and to treat its members as equal partners.

Lesseps did not agree. On his return to Cairo, he wrote to Arlès-Dufour knowing that everything he communicated would in turn be relayed to Enfantin. Lesseps admitted that he too was worried about the truculence of Stephenson, and for that reason, he thought it best not to emphasize the role of the Study Group. If one of its key members, a person allied with the governing coalition in England, was actively opposed to the canal, that would make it almost impossible for the group to function effectively. Besides, Lesseps continued, it was a moot point, since Said had entrusted the building of the canal not to a group but to one man, Ferdinand de Lesseps. That was for the best, he continued, because the Turks preferred personal relationships, and placing a project in the hands of someone trusted carried more weight with Said than a multinational group, no matter how impressive its pedigree. He told Arlès-Dufour that he planned to remain "the master of this affair." He acknowledged that the Study Group had been a vital element, but it was his opinion that the group "no longer exists and there is no way to revive it."

This was a debatable conclusion. The group had been dormant, yes, but it had never officially disbanded, and its leading members contended that it did still exist. Lesseps did not argue the merits of what the group had accomplished, and he credited Enfantin, Arlès-Dufour, the Talabots, and Negrelli with having done invaluable work. For that reason, he explained, their names had been included as founding members of the Suez Canal Company. In financial matters he was equitable, and had every intention of remunerating the Study Group with whatever profits accrued from the canal. Yet, though he wanted its members to be rewarded for their efforts, he was determined that they have no say in how the project would evolve from this moment on.

Democratic about sharing the wealth, he was autocratic when it came to personal glory. That desire was the key ingredient of his ambition. As he told his mother-in-law several days after writing Arlès-Dufour, "I want to do something great, without hidden motive, and without any personal interest in money. . . . I will be resolute. . . . My ambition, I swear, is to be the *only* one to lead all those involved in this immense affair. . . ." He refused to accept any conditions that anyone placed on him or on his vision for the canal. He claimed that he had learned a crucial lesson from none other than Muhammad Ali, who had once told him that if he wanted to accomplish anything important he could only count on himself. "If there are two of you," the pasha had advised him, "there are too many of you." Lesseps took that lesson to heart, and it was his guiding mantra for the remainder of his life.[10]

Enfantin did not react to this snub with equanimity. Surrounded by acolytes who viewed him as the messiah, he was not used to having his wishes so flagrantly disregarded. He began a campaign to undermine Lesseps. In France, he assailed both the logistics of the proposed plan and the validity of Lesseps's claim. He enlisted supporters to spread the word that the concession was illegitimate, that the direct route violated both common sense and good engineering, and that Ferdinand de Lesseps, a once-disgraced diplomat, had usurped the venture in ways not befitting a gentleman.

He also wrote directly to Lesseps. His letter was alternately taunting and conciliatory. He accused Ferdinand of unnecessary paranoia that others might steal his thunder, and stated that such fears were not becoming in someone with pretensions to greatness. He scolded Lesseps for relying on the advice of Linant, whom Enfantin had tangled with in years past and whom he now accused of a complete lack of scientific knowledge. He also questioned Lesseps's technical acumen. "We [Saint-Simonians] are engineers," he chided. "Are you and Said engineers?" Furthermore, he assailed Lesseps for not seeing how much of a liability Linant was. The English opponents of the scheme were using Linant's plan and the Pelusium idea as the major point of attack. If Lesseps would see the light, embrace the support of Enfantin, and adopt the ideas of Talabot for an indirect route, then the canal might stand a chance of being built. Otherwise, Enfantin warned, the project "would find not a sou in Europe for an enterprise so obviously crazy."

Finally, Enfantin wanted to know how Lesseps could dismiss the Study Group but make its members founding agents of the company. How could they be founders if Lesseps refused any input? How could they be part of a venture in which they had no say? Perhaps fearing that he had left no room for compromise, Enfantin ended his letter on a measured note and politely asked Lesseps to keep him informed and not to exclude him from these serious deliberations.[11]

The gloves now off, Lesseps wrote more forcefully to Arlès-Dufour. Like a juggler, he had to balance multiple concerns, and the opposition of an eccentric millionaire in France was not one of the concerns he really wished to spend time on. He didn't understand what claim the Study Group could have, especially since Said was now paying for the expense of the new surveys of the isthmus and for Lesseps's travels. He corresponded with Arlès-Dufour once in mid-February, on the day that he arrived in Constantinople, and then ignored the Saint-Simonians until June. The fact that he had gone to Constantinople before first returning to Paris offended Enfantin. He expected Lesseps to return and consult with him before moving ahead. But Lesseps decided that, though Enfantin's support was desirable, it was hardly essential. Returning to Paris would consume valuable time, and Lesseps worried that unless he obtained the sanction, or at least the partial sanction, of the Ottomans his ability to control the project would be compromised. Either he went to Constantinople to state his case, or his opponents would state it for him.

As Lesseps entered the labyrinthine world of European diplomacy during the spring, Enfantin continued to fulminate, and Arlès-Dufour wrote Lesseps a series of letters imploring him to respond. Enfantin was able to arrange another interview with the emperor, who he claimed was sympathetic to his case. But Napoleon III would often agree with those meeting him rather than risk the discomfort of con-tradicting them, and he made no move to impede Lesseps. As it became clear to Enfantin that his opposition was making little head-way, his letters to his supporters became more shrill. He saved some of his best vitriol for Linant, whom he called "personally monstrous, ego-tistical, and capable of any sacrifice to sate his puerile vanity." He lamented that, though Lesseps had swooped down and appropriated "the old and great dream" of the Saint-Simonians, Lesseps himself "has dreamed far too little about these affairs and has ignored diplomatic, financial, and artistic precedents."[12]

The financial precedents were the moneys invested by the Study Group; the artistic were Talabot's plans for an indirect route; and the diplomatic were Lesseps's decision to go first to Constantinople and then not to pay sufficient attention to the problem of England. Yet Lesseps proceeded as if oblivious to Enfantin's qualms. This inatten-tion infuriated the proud Father, and he vented about Lesseps in a rambling letter: "We want to eclipse him, make him disappear, subor-dinate him, and rob him." He predicted that Ferdinand would soon fall, "just as he did in Rome in 1848! If he thinks that having Said as an ally will help him, he's mistaken." Enfantin continued, "Lesseps will come to see that to be the door warden of Said . . . is without prestige among men of state and affairs."

Enfantin's opposition continued throughout the year; he made a final stab at sabotaging Lesseps in the fall, but Lesseps was no longer fighting back. As of February, he was focused on more immediate obstacles. He wrote to Arlès-Dufour one last time in June, expressing astonishment that the Study Group was still objecting. Arlès-Dufour had written a plaintive letter claiming that he had never ceased to be Lesseps's friend and ally, but Ferdinand did not see it that way. When it came to the Suez Canal, he took Muhammad Ali's maxim to heart: two were too many, and working with Enfantin and the Saint-Simonians was not acceptable. Though he claimed he had nothing against Arlès-Dufour personally, it is clear from this last letter that, in his mind, one was either for him or against him, and that Arlès-Dufour could not be both an ally of Lesseps and a follower of Enfantin.[13]

Lesseps showed that he could abruptly terminate a friendship if he perceived the slightest disloyalty. He also proved an astute judge of which battles required his attention and which did not. In the coming years, he rationed his energy by attending only to those crises that threatened to submerge the project. By the spring of 1855, the dispute with the Study Group did not pose such a threat, and so he ignored it. English opposition to the canal, and the British influence in Constantinople, however, could have been lethal. Having barely celebrated the victory of winning the concession from Said, Lesseps turned to the formidable challenge posed by the British Empire and its ambitions for the Middle East and the world.

# CHAPTER EIGHT

# THE SULTAN'S SHADOW
# AND THE ENGLISH LION

A RRIVING IN CONSTANTINOPLE in February 1855, Lesseps arranged an audience with the British ambassador. Constantinople was the capital of the Ottoman Empire. Before becoming the seat of Turkish power, which stretched from Vienna to Persia, it had succeeded Rome in the eastern Mediterranean, and the Byzantine emperor had governed the temporal and spiritual lives of millions. But by the middle of the nineteenth century, the British ambassador was arguably the most important official there.

Lesseps also hoped to meet with the sultan, the grand vizier, and the foreign minister. The vizier was the equivalent of a European prime minister. While the sultan shuttled between the Topkapi Palace complex in the old city and the nearly completed Dolmabahce Palace on the Bosporus, the vizier and other members of the government were located in a small palace named the Sublime Porte, in honor of its elaborately decorated gateway. Just as the U.S. government is often referred to simply as "Washington" and the British government as "Whitehall," the Ottoman government was usually called "the Porte," after the building where its primary offices were located.

The Ottomans didn't care deeply about whether the canal was built. They were concerned about the political ramifications. If the construction of the canal could be used to increase the power of the decaying empire, the Ottomans would back it; if it appeared that the canal might further weaken their influence, they would resist. Neither position was entirely satisfactory. The canal could revive the commerce of the eastern Mediterranean and thereby strengthen the Ottomans, but it could also enhance Egypt and turn Said into a dangerous competitor. Preventing construction might placate the British, who would then be more inclined to defend the integrity of the Ottoman Empire. But resistance might also alienate the French, who had been loyal and much-needed allies. The only certainty for the sultan and his ministers was that Russia was an enemy who would annihilate the Ottomans unless Britain or France prevented it.

It was a humbling period for the Ottomans. In 1453, Sultan Mehmed the Conqueror captured Constantinople and ended the Byzantine Empire. Though Byzantium had been declining for centuries, the fall of the last bastion of imperial Rome by a Turkish dynasty was a wrenching moment in Western history. Europe was deeply shaken. The Ottoman conquest did not stop. In subsequent decades, the armies of the sultan pushed into Eastern Europe, and annexed what are now Bulgaria, Romania, Yugoslavia, and Hungary. The Ottomans kept advancing, all the way to the gates of Vienna in 1529. The siege was unsuccessful, and Sultan Suleiman the Magnificent ordered his troops to return home before the onset of winter. But the empire did not recede. It expanded, throughout North Africa, into the Near East as far as Persia, and around the Black Sea, through Tartary, and into the Caucasus, where the Ottomans perennially defeated the soldiers of the Russian tsar.

For the next two centuries, the Ottomans threatened Europe. They had a reputation for brutality, though they were no more violent than the Hungarians, the Spanish, or the other European powers that they confronted. The only difference was that the Ottomans usually won, and thus were able to slaughter and enslave rather than be slaughtered or enslaved. Having occupied Constantinople, the greatest capital of Christendom, they made it their own, and built mosques to rival the churches of the Byzantines. They also ruled the Arabian Peninsula, and the sultan enjoyed the title of protector of the holy cities of Mecca and Medina. The Ottomans constructed a fleet to challenge the Venetians and the Spanish for supremacy in the Mediterranean, and although the Spanish naval victory at the battle of Lepanto in 1571 was heralded in the West as a turning point, the empire was so rich that it soon replaced the lost ships. Not until 1683, when a Polish army saved the desperate Austrian defenders at the gates of Vienna, did the

Ottomans cease to challenge Europe. And not until the end of the eighteenth century did the Russians master the Ottomans in battle and begin to reverse centuries of expansion.

In the first decades of the nineteenth century, after devastating military and territorial losses to the Russians, the sultan realized that changes were necessary. The old military system of slave Janissaries had become corrupt and ineffective, and in 1826 Mahmud II emulated Muhammad Ali and slaughtered the Janissaries after they had rebelled against him. He then built a new military, modeled on the armies of Europe. But he was unable to modernize as quickly as his Egyptian vassal, and without the intervention of England, he would have lost his throne to the pasha. Mahmud died before the crisis of 1839 was resolved, and left his inexperienced son, Abdul Mejid, to deal with the Egyptian invasion, the Europeans, and a declining empire.

Having been saved by the British, Abdul Mejid and his ministers could no longer deny that they were living on borrowed time. The only thing preventing the Europeans from dismantling the empire was their inability to agree on who would get what. The British didn't want the Russians to control Constantinople and the Dardanelles; the French wanted to keep the British from becoming the pre-eminent power in the eastern Mediterranean; and the Russians wanted to prevent the Austrians from annexing the Balkans. Unable to defend itself but kept alive by the rivalries of its adversaries, the Ottoman Empire became known as "the Sick Man of Europe."

Like Said, Sultan Abdul Mejid lacked grit, cared deeply about the empire's fate, and followed a reformist path charted by his strong-willed father. But, unlike Said, he was blessed with a group of dynamic reformers who tried to end corruption and untangle the archaic bureaucracy. By the 1850s, imperial policies were largely determined by a triumvirate of high officials. The most influential was Mustafa Reshid Pasha, who before becoming grand vizier had served as ambassador to Paris and spoke fluent French. Along with his younger colleagues and sometime adversaries, Ali Pasha and Fuad Pasha, he was an advocate of a radical reorganization of the Ottoman bureaucracy along European lines. He wanted to transform the sultanate into an enlightened monarchy under a constitution and answerable to the rule of law. Traditionally, the Ottoman state had been a pyramid topped by an absolute monarch who not only had the power of life and death but also acted as both king and pope. His word was law, and religious authorities deferred to him. Now the empire was in flux; tradition was being assaulted from within and without; and the last thing the Ottoman government wanted was a Suez Canal that would upset matters still further.

Lesseps moved fluidly in the diplomatic circles of Constantinople. He was accorded every courtesy, both in deference to his former position in the French Foreign Ministry and to his current role as the Egyptian viceroy's emissary. But although he was entertained royally and treated as a visiting dignitary, the more he talked with Ottoman officials, the clearer it became that without the blessing of the British the sultan was not prepared to sanction the canal. Lesseps had hoped that the road to Suez went through Constantinople. He discovered instead that it went through London.

Even if the Ottoman Empire was a shadow of its former self, it was not without resources, one of which was a legacy of diplomatic intrigues with the Europeans. The men who led the empire in these years understood that they could survive only by keeping European expansion at bay, and the only way to do that was to play one state off another. It was a position of weakness, but the Ottomans used it adroitly.

The relationship between the sultan's government and the British Empire was the key to Ottoman survival. Having created the world's most powerful navy, the British expanded throughout the world in the nineteenth century. When British officials contemplated their strategy, they focused on vital sea-lanes: the English Channel, the Strait of Gibraltar, the Cape of Good Hope, the North Atlantic route to America, the Dardanelles and the Bosporus. Direct control was not necessarily required, but none of those sea-lanes could be allowed to fall to an adversary or competitor, and in the middle of the nineteenth century, the British treated all European states as potential competitors. British imperial strategy, therefore, dictated that there were only two viable options for the Bosporus and the Dardanelles: direct British rule, which would arouse the animosity of the Russians and the French and probably involve the English in a long, costly war, or neutrality under Ottoman sovereignty.

To maintain the Ottoman Empire and undermine the French, England fought Muhammad Ali in the 1830s. Using similar calculations, Britain opposed the Suez Canal in the 1850s. The English governing class was drawn from a small community of aristocrats in these years, though democracy was starting to encroach on the privileges of the landed elite. The Reform Act had been passed in 1832, significantly increasing the number of men eligible to vote, and the rise of industry had shifted power away from the traditional estates of the countryside and toward the cities of London, Glasgow, Manchester, and Birmingham. Even so, in the 1850s and 1860s, power rotated between a small number of powerful men; few were more powerful than Lord Palmerston, and few diplomats had more influence than the imperious British representative to the Sublime Porte, Lord Stratford Canning de Redcliffe.

Palmerston was a dominant force in English politics for almost fifty years. Graced with a mop of wild hair and bushy sideburns, he was the very model of a particular sort of English gentleman. He was not the genial father of a Jane Austen novel but rather a stern figure, watchful of the perils that lay on the other side of the English Channel. Among friends, he had a reputation for generosity and good humor, but those qualities were not evident in public. In portraits, his face usually carried an expression of dyspeptic hauteur. He was foreign minister for much of the 1830s and 1840s, during which time he created the mold of the unyielding, jingoistic English bully threatening the world. He was also the single greatest obstacle to the construction of the Suez Canal. An unrepentant Whig, Palmerston was prime minister from February 1855 till 1858, and again from 1859 until his death in 1865. And at no point in that entire time did he soften his opposition to Lesseps.

"His heart always beat for the honour of England," Lord Russell eulogized in 1865. Palmerston believed in England and he believed in the British Empire. Whether he believed deeply in anything else is open to question. That set him apart from some of his younger rivals. William Gladstone had a missionary zeal for Christianity; Benjamin Disraeli flirted with writing novels and becoming a man of the mind. Palmerston, however, was not distracted by alternate life-paths. To call him imperious is to understate the contempt he held for much of the non-English world. He rarely uttered a public or private word suggesting self-doubt, and he was not known to concede that there might, perhaps, be more than one side to a story. For Palmerston, there was only one side—his, or, rather, England's—and it was always the side of right.

These qualities undoubtedly gave him an edge in politics, both at home and abroad. His certainty allowed him to take swift action in times of crisis. Because he always knew where he stood, the proper course was always evident to him. He had one overarching morality: the health of the British Empire and the sanctity of the English way of life. He expected the rest of the world to act in its own interests; he viewed history not as an expanding pie of happiness, but as a Hobbesian struggle for a limited amount of global wealth and power. Because no one would protect British interests except the British crown, it was his duty as Her Majesty's representative to guard the empire. Whether that pleased or displeased the other states of the world was not his concern. Under his watch, Great Britain would have no allies, and its policies would be determined by Her Majesty's Government alone.[1]

Palmerston was unusually adamant and extraordinarily consistent, but he reflected widely held beliefs in nineteenth-century England, Scotland, and Wales. The British had been building an empire for centuries, but only in the mid-nineteenth century did they ascend dramatically above the rest of Europe. The late eighteenth century had seen setbacks for the British in North America, and then a prolonged period of war and uncertainty while Napoleon threatened to transform the European continent into a French province. Palmerston was part of a coterie of high officials who championed the idea of Britain alone and Britain supreme. The strategy worked. By the 1850s, the English navy ruled the waves, and English armies governed India, Hong Kong, Australia, New Zealand, and, to a lesser extent, Canada. It was not territory that defined British power, however, but trade and industry, and by the time Palmerston became prime minister, both were burgeoning.

The British were the pre-eminent trading power in the world, and during the years when the Suez Canal was being built, Britain's economic might increased significantly. Between 1850 and 1870, British exports tripled, from just over eighty million pounds to more than 240 million pounds a year. The process was fairly straightforward. The British imported raw materials from every corner of the globe. They then used those raw materials, transformed them into finished products in factories, and exported those goods throughout the world. Trade and industry were inextricably linked. The United Kingdom needed raw materials to produce finished goods, and it needed markets to absorb those goods abroad. In order to profit from exports, it had to control the trade, and to do that, it had to control the seas. In that sense, the British navy was simply an adjunct to the British merchant marine.

More than any other country at midcentury, Britain was a manufac-
turing nation. Factories sprang up in the Midlands, and coal from
Wales and Newcastle helped fuel this new industrial behemoth that
produced steel for railways, turned jute from India or cotton from the
United States and Egypt into clothing, and then sold those finished
products at a healthy profit in India, Argentina, China, and Europe.
Soon, other countries would begin to emulate the British model. But
whereas the United States could industrialize with domestic raw materials, Britain did not possess sufficient resources. It had to trade in
order to support its industrial economy, and in the world of competing
states, that required an aggressive, independent foreign policy and the
means to enforce it. For Palmerston and for British society, command
of the seas was the oxygen that the empire breathed.[2]

In France, industry and progress were cultural mantras for much of
the middle class. In England, these notions were somewhat offset by a
streak of pessimism. It wasn't that the English believed any less fervently in the cult of progress, but there was a sense that it could all go
terribly awry at any moment if the wrong policies were pursued. Too little democracy, and power would corrupt; too much, and chaos would
ensue. In literature, popular culture, and newspapers, and in the
speeches and statements of leading politicians and businessmen, the
British exuded a belief that their empire was destined to expand, that
their moral purity was unparalleled, and that liberal virtues gave the
empire strength and fortitude. But there was also a current of unease.
Queen Victoria could exclaim, at the opening of the Crystal Palace
Exhibition in London in 1851, that under one vast roof "was displayed
all that is useful or beautiful in nature or in art." The hand of God, she
continued, was visible in the many wonders of human artifice assembled there. Britain was blessed among all nations, and the future beck-
oned with unimagined wonders. And yet, in the midst of this
celebration, many questioned the materialism of the empire and worried that it would be England's undoing. The poet Matthew Arnold
considered affluence a weakness. All the innovations of the modern
age, he concluded sadly, formed a flimsy defense against the dragons of
anarchy.[3]

Not fully trusting human nature, Palmerston was skeptical of democracy and tried to keep the reform movement from consolidating
the gains it had made in the 1830s. He opposed the liberals of "the

Manchester school," who held that free trade was an unalloyed good. He did not believe that trade without import duties would strengthen the empire, and thought instead that it would flood Great Britain with cheap manufactures and destroy the foundations of the economy. He distrusted foreign advocates of free trade even more. A foreigner who spoke of liberalism, free trade, and progress was simply dressing up competition and animosity in pretty words. And when Ferdinand de Lesseps and the partisans of the Suez Canal claimed that its construction would benefit not just France and Egypt, but England especially, Palmerston suspected a forked tongue and dark motives.

His attitude was shared by Lord Stratford Canning de Redcliffe in Constantinople. Stratford was a force of nature in his own right, and, like every other British foreign-policy official in these years, he opposed the Suez Canal from the moment he heard about it. Like Palmerston, in portraits he often seemed to be scowling, and in his diplomacy he was fierce and intimidating. Carrying the confidence of Her Majesty's Government, he was a respected and feared presence in Constantinople. His influence was so great that he was known, and not affectionately, as Sultan Stratford or Abdul-Canning.

Stratford perceived himself the arbiter of the affairs of the Ottoman Empire. The sultan and his officials saw things differently, but they knew that their best interests lay in stroking Stratford's vanity. They respected his authority and the power of Great Britain, but they viewed the British as one of several pieces of the strategic puzzle. In the spring of 1855, the Ottomans allied themselves with the British and the French against the Russians in the Crimean War, named for the Black Sea peninsula where much of the fighting took place. Between the defeat of Napoleon in 1815 and the outbreak of World War I in 1914, the Crimean War was the only time the powers of Europe engaged in long and protracted confrontation, and it was a conflict without strong passions or clear reasons. Ostensibly, the war was caused by Russian and French competition over the Holy Lands. Both nations vied to be the official protectors of the Christians who lived in Jerusalem. But that was only one spark. Russian attempts to exert control over the Slavic Christians of the Balkans angered the Ottomans. The British, for their part, sought to limit Russian naval power. Even though these were all issues of concern, they did not justify the carnage that ensued. The Crimean War did give Western civilization the charge of the Light

Brigade, the evolution of the Red Cross as Florence Nightingale tended to the troops in Scutari, and a biting George Bernard Shaw satire. It also went down as a war of folly and unnecessary bloodshed.

While Britain and France may have temporarily allied with each other and with the Ottomans, that didn't mean there were strong or lasting ties between the three. Stratford was as wary of the French as he had always been, and his colleagues in London concurred. The Ottomans understood that neither the British nor the French would hesitate to occupy Constantinople if they could, and that the alliance was purely the result of European balance-of-power politics. That the siege of Sebastopol was being run ineptly only added to the tension.

Lesseps would have faced an uphill struggle regardless of the situation in the world at large, but the Crimean War did not make his task any easier. Yet, though the British government of Lord Aberdeen had just fallen and been replaced by Palmerston, and though Constantinople was rife with speculation about the future course of the war, Lesseps found it relatively simple to arrange meetings. This was partly because he arrived with a letter from Said asking that the bearer be treated as the viceroy's emissary; but, more to the point, Lesseps himself had the respect of the diplomatic corps. He had been one of them, and they had interpreted his disgrace in 1848 as one of those unfortunate moments when a diplomat is made a scapegoat for the failings of generals and princes.

However, though he was welcomed in the salons of Constantinople, neither the British nor the Ottoman officials planned to give him what he wanted. Stratford had both strategic and personal reasons for opposing the canal. Personally, he was connected to factions involved in railroad projects, and he favored Robert Stephenson and the railway between Alexandria and Cairo. Strategically, he saw the railroad as a route that could be controlled by Britain, and he considered the canal project a French scheme that would enhance Napoleon III. He also doubted that Lesseps would be able to act on his own as a lone entrepreneur. Like Frederick Bruce, the consul general in Egypt, he believed that the French government would eventually throw its support to the canal, and its completion would become a matter of French honor. Stratford hoped to kill the idea before it developed momentum.[4]

But he worried that Lesseps would persuade the Porte to countenance the plan. The Ottomans had no vested interest one way or the other, and the sultan would be loath to offend the French. To his frustration, Stratford was also constrained in what he could say. He could insinuate that the project presented multiple problems, but he had been instructed by his government not to object officially. As antagonistic as the British were, Palmerston wasn't prepared to jeopardize the alliance with France. But, while avoiding direct confrontation, the British concluded that it wasn't necessary to take a public stance against the canal. All they had to do was make sure that the Ottomans did not formally permit Said to go forward.

Within days of his arrival, Lesseps recognized the bind that he was in. The Ottomans were annoyed at Said for granting the concession without consultation; the British were determined to make the road arduous without provoking a diplomatic crisis; and the French were not willing to create an incident with the British over what amounted to an overly ambitious idea by a private French citizen.

Lesseps had come looking for a decision that no one was willing to make. In response, he personalized the contest. In dealing with the Ottomans, he contended that the viceroy did not need the permission of the sultan to grant a concession for the canal, any more than he had to have permission for the other large-scale public-works projects that had been carried out in past years. Lesseps acknowledged that a blessing from the Porte was desirable, but he never accepted that it was legally required. He claimed instead that the canal was a contract between Said and himself, and that was all he needed to form the company and begin the canal.

In his dealings with the British government, Lesseps began to argue that the opposition of Stratford and Palmerston did not represent either British public opinion or, for that matter, Britain's best interests. Dismissing the attitude of Stratford as a "personal objection" rather than a legitimate reflection of the stance of the British, Lesseps asserted that Britain, as the dominant force in world trade, would gain the most from the canal's construction. He also strongly hinted in his meetings with Stratford that it would be foolish to stand in the way, since it was only a matter of time before the sultan approved. He could not have been sure of that, of course, but the bluster unsettled Stratford and, by extension, Palmerston.[5]

Both men became convinced that Lesseps harbored "ulterior motives." Some of their paranoia was merited. As Lesseps wrote to an ally in March, "England does not like to confess the motives for her opposition; but she must be persuaded that she can no longer monopolize the commerce of the world, or the domination of all the seas." He may have believed that the canal would benefit everyone, but he did not mind if it increased French power at the expense of the British.

Even while sparring with the British, Lesseps was thinking ahead to the eventual shape of the company and to how the venture might be financed. He had already committed himself to the idea of a company owned by shareholders throughout Europe and Egypt. More specifically, he thought that it should be owned not by a few wealthy individuals but, rather, by "the greatest possible number of small shareholders." Otherwise, the large bankers of Europe would demand a voice in the management of the company and the construction of the canal, and Lesseps was no more willing to cede control to bankers than he was to Enfantin and the engineers of the Study Group. Though there was little precedent for funding a project of this scale by the flotation of shares to thousands of people scattered throughout two continents, Lesseps proceeded with the same blithe confidence in financial matters that he had in engineering. Without a deep understanding of either, he was able to arrive at a course of action that eluded those who knew better.[6]

During the next months, Lesseps traveled constantly. Indeed, that was to be his pattern for the next fourteen years. The diplomatic template established in the spring of 1855 did not change significantly, and the British only ceased their relentless opposition when Palmerston died in 1865. During the intervening years, however, the situation in Egypt, Constantinople, and France changed substantially, sometimes to the advantage of Lesseps and the *canalistes*, sometimes not.

Lesseps left Constantinople without any tangible results, except that he was now more aware of the obstacles in his way. Meanwhile, his visit had widened the fissure between Said and the sultan, as well as between Said and those who opposed the canal in Egypt. The merchants of Alexandria organized to prevent the canal from being built. They were getting rich from increased trade with Europe, and the completion of the railroad promised more riches in the future. The canal, especially a direct canal through the isthmus, would bypass them. Several members of Said's court had close ties to the merchant community, including Kiamil Pasha, his brother-in-law. Kiamil was whispering that the canal was a bad idea. He then wrote to Said stating his objection, which led the viceroy to return a sharp response. "Your letter," Said replied, "would imply that I inclined toward the French, and you think that the canal originated in my wish to do something agreeable to them; while at the same time you confirm the statement that you are all in great fear and mortal terror of Lord Stratford de Redcliffe and the English." He then concluded with a staunch defense of his actions and his integrity. "I am convinced of the great and special advantages of the undertaking for Islam and for Egypt itself, and I have behaved in this matter like a good Turk."[7]

Said rarely invoked Islam. It was understood that he was the ruler of Egypt and, as such, the defender of Egyptian Islam in much the same way that the British crown was the defender of Anglicanism. Yet religion was hardly ever a factor in his decisions, and the days when Middle Eastern societies would be challenged by fundamentalists lay far in the future. There was no formal separation of church and state because there was no church, and Muslim clerics did not meddle with affairs of state. Neither Said nor his father was especially devout, and though Said liked the idea of improving the position of Muslims in world affairs, he was driven by dynastic interests rather than by the Koran. Whether or not he worried about his standing in the eyes of God, he cared deeply whether the Turkish elite of Egypt believed he was conducting himself with the honor befitting a Turkish prince. As stern as he was in response to Kiamil, the domestic opposition troubled him, and over the coming months his initial enthusiasm for the canal began to dissipate. Lesseps had spun a fantasy that was proving much more complicated in reality.

Lesseps, recognizing that Said's ardor was cooling, spent several weeks in Egypt working on plans with Linant and Mougel and providing Said with moral support. But he was eager to return to France and then go to England. Hearing of Enfantin's attempts to get the emperor to stop the project, Lesseps felt it was important to make his own case in person. With the help of the empress, Ferdinand trusted that he would be granted an audience so that he could answer Enfantin's objections.

He was not particularly worried. Having studied the career of Louis-Napoleon, Lesseps concluded that the emperor would ultimately support the canal. In letters to his mother-in-law, he talked of how Louis-Napoleon, while still in prison and then in exile, had developed a fascination with canals, especially with a canal through Nicaragua to connect the Atlantic and the Pacific. The emperor had even written a memo about the virtues of such a waterway, in which he said: "War and commerce have civilized the world. Commerce is still following up its conquests. Let us open up a new route for it." Lesseps was confident that Napoleon would see Suez in the same light.[8]

In March and April, two articles appeared in the influential *Revue des deux mondes*. The journal covered topics of interest throughout the world, with essays on everything from the American circus man P. T. Barnum to the mores of the Bedouins of Syria. The two articles portrayed Lesseps as a fool and the direct route as misconceived. One, by Paulin Talabot, reiterated the stance of the Study Group, and stressed that Lesseps was a neophyte who refused to learn from the wisdom of past canal builders. Talabot used the example of the ancient Ptolemies as a rebuke: previous canals had used Nile water and not sea water, yet the direct route would force a mingling of two different seas, with different currents and tides. The result, warned Talabot, could be flooding and rapid erosion. The other article, by J. J. Baude, was wildly positive about a Suez Canal, but not about Lesseps's plan. Yes, Baude acknowledged, the canal was a magnificent idea, one that would enrich the French Empire in Algeria and beyond and even revitalize poor Venice, which had been so sadly eclipsed by the route around the Cape of Good Hope in the sixteenth century. But, Baude cautioned, the canal was an international issue that would affect every major state and several minor ones. It was not appropriate for a lone individual such as Lesseps to arrogate to himself the power to decide the fate of millions. A canal would have universal consequences, and the decisions about its shape and scope should be made by the powers of Europe in consultation with the many private and public parties who would be affected.[9]

These attacks criticized Lesseps, but at least they supported construction. That was half the battle. The British, however, were opposed to any canal, not just the one proposed by Lesseps. Before leaving for England in June, Lesseps returned to France and was finally permitted to meet with the emperor. Officially, Napoleon demurred. He told Lesseps that now was not the time to confront the British and that the Crimean War and considerations of European politics made it unwise to risk a breach over what still amounted to a scheme. He urged Lesseps to take his case directly to the British, and warned that the active opposition of London was not something that he as ruler of France wished to challenge.

That was Napoleon's stated position. According to Lesseps, however, Eugénie privately broached the subject with her husband. The emperor told her that he thought that the canal was a wonderful idea, and that he hoped that Lesseps would succeed. His public demurral, he assured her, did not reflect his personal opinion, and he said that though he could not actively lend his support, he certainly would do nothing to prevent the project from moving forward. Relieved that he would not face problems in Paris, Lesseps went to London to confront Palmerston directly.[10]

# CHAPTER NINE

# HITHER AND YON

L ORD PALMERSTON WAS prime minister of the most powerful
country in the world in June 1855, yet, within days of arriving in
London, Lesseps had an invitation to call at the viscount's home.
True, the world was a smaller place in the middle of the nineteenth
century. The circles of influence and power consisted of close-knit
groups, and though gaining entry was difficult, meeting others once
entry had been gained was not. Still, even by the standards of the time,
Lesseps had an uncanny ability to see whomever he needed whenever
he needed to. And that summer in London, it was Palmerston's turn to
be seduced.

Much to Lesseps's dismay, Palmerston proved immune. Not for the
last time, a passionate Frenchman found that his charm was lost on the
English. Lesseps came armed with the same arguments that had
proved so persuasive in France and with Said, but the reaction in England was skeptical. Palmerston had always viewed the French as
prone to unrealistic dreams that created havoc for their neighbors.
Though he did not think that Lesseps was as dangerous as Robespierre
or Napoleon, he still saw the canal as the latest in a series of bad
French ideas that were best left unrealized.

Palmerston enjoyed the quiet opulence of a mid-Victorian gentleman, but the London of 1855 was a teeming, messy, filthy place of more
than two and a half million people and growing rapidly. The pristine
central districts were juxtaposed with new neighborhoods built to
accommodate the immigrants from the countryside who worked the
factories or who serviced the increasingly complicated economy. Transportation was chaotic. The main streets were insufficiently wide and
often jammed as people made their way on foot, on animals, in car-

riages, or in a new form of mass transit called the horse-bus. Smelly pubs swimming in cheap ale coexisted with sedate salons offering a servant for every guest.

High culture was shaped by the ideas of men such as John Stuart Mill, who called for a liberal, limited government that served the needs of the many and protected the rights of the few; it was flavored by the Romanticism of critics like John Ruskin, who extolled the virtues of nature and the moral responsibilities of art; and it bore the stamp of Queen Victoria and her consort, Prince Albert, who represented a detached monarchy atop a well-ordered society.

Yet, however well ordered in theory, it was anything but in reality. London was equally shaped by its merchants and clerks and workers who tried to carve out their own niches and profit from the industry and commerce that converged in the city. They devoured the serialized novels of Charles Dickens in the weekly magazines that proliferated along with the growing population and rising literacy rates. Reveling in Dickensian tales of young boys making their way through the labyrinthine obstacles that the city presented, these groups were more likely to poke fun at the pretensions of the aristocracy and demand a greater say in the affairs of the country than they were to genuflect to traditional notions of the landed gentry and rule by the few. Palmerston enjoyed their support when he defended British honor abroad, but he was ever more out of step with the industrial city and the industrializing country whose government he led.

Many of these tensions and changes are more apparent today than they were at the time, but Lesseps was certainly aware that Palmerston was both a powerful man and an anachronism. Though he hoped to change the prime minister's mind, he was more confident that, once the virtues of the canal were explained to the British people, they would embrace it, and the government would then follow suit. It was not that simple. Eventually, a significant portion of Britain did support the construction of the canal, but never with the unbridled enthusiasm of the French. In Britain, the canal was from the outset indelibly perceived as a French waterway, rather than a universal venture for the benefit of all. Lesseps preached the canal's merits, but he never fully appreciated the depth and breadth of British concerns that it was all a sinister French plot.

The first encounter between Palmerston and Lesseps was nothing less than civil. Voices were not raised; harsh words were not exchanged.

But the lines were drawn nonetheless. According to Lesseps's account of the meeting, Palmerston was blunt about his reservations. "I do not hesitate to tell you what my apprehensions are," the prime minister said. Palmerston didn't think that the canal was technically viable, and even if the engineering challenges could somehow be overcome, he felt that the opening of a new route to the East would undermine England's position as the dominant power in world trade. The Suez Canal would be available to all countries, and could therefore decrease the importance of English merchants. He also questioned the motives of the French. Relations between the two countries might currently be warm, but there was no guarantee that they would be in the future, and if hostilities erupted, a Suez Canal incorporated in Paris would be a distinct advantage for France.

As Lesseps listened, he realized that there was nothing he could say to sway Palmerston that morning. But he asked the prime minister to consider the advantages that a canal would offer Great Britain. It would shorten the route to India and the East by thousands of miles, and even if, by some unforeseen chance, England and France found themselves at war, the British navy would still enjoy immense advantages from the canal, which the French would be hard-pressed to circumscribe. As for the engineering challenges, Lesseps assured Palmerston that they were not insurmountable and that an international commission of engineers would shortly be dispatched in order to prove once and for all that the canal and the jetties planned for Port Said were feasible.[1]

Over the next four years, until the formation of the Suez Canal Company in the fall of 1858, Lesseps met with Palmerston several more times. The two continued to spar over the canal, but the arguments never departed dramatically from the battle lines laid down at this first encounter. Palmerston became increasingly shrill in his attacks—not when confronted with Lesseps in person, but in his statements during parliamentary debates and in various other forums, public and private. Lesseps gradually concluded that the aging lord had lost touch with common sense, and that his opposition to the canal was based not on reason but on paranoia. In essence, their attitudes did not evolve from that morning in the spring of 1855. Lesseps believed that the canal would benefit the world and ultimately do no damage to British security; Palmerston was convinced that the opposite was true.

If anything, the viscount had pulled his punches during their first discussion. A week later, Lord Cowley, the British ambassador in Paris, expressed his own view of what the position of Her Majesty's Government should be. Writing to the foreign minister, Cowley said that the French for years had sought to undercut the British railway in Egypt and supplant British influence by digging a French canal. Lesseps's arguments were pure fantasy, and asking for British support was absurd. "It would be," he concluded, "a suicidal act on the part of England to assent to the construction of this canal."[2]

That was the sentiment of the British diplomatic corps and of Palmerston. But for political reasons, British officials refrained from stating their stance so baldly. Doing so might jeopardize peaceful relations with France, and the English were as perplexed as the Ottomans about Lesseps's relationship with Emperor Napoleon. Would a rebuff of Lesseps be the same as rebuffing the ruler of France? That was not clear, and in the absence of conclusive proof that Lesseps was not an agent of Napoleon's will, it was prudent not to risk an international crisis. The British government, therefore, much like the Ottomans, adopted a policy of passive resistance. Palmerston and his ministers always raised questions and expressed reservations. But, aside from occasional intemperate outbursts by Palmerston as the years went on, they stopped short of unequivocal opposition.

That did not prevent their allies in the press and in Parliament from using unequivocal language, however. In preparation for his public-relations campaign in England, Lesseps published a widely distributed pamphlet called *The Isthmus of Suez Question*. The basic argument was the same that Lesseps had been making in Cairo and Constantinople. He praised the virtues of a neutral ship canal that would augment international trade by shortening the route to the East from the more than eleven thousand miles around the Cape of Good Hope to just over six thousand miles. He claimed that the canal could be constructed in slightly more than six years, at a cost of between six and seven million pounds. Though this was expensive, he pointed out that it was half the amount spent on the rail line between London and the northern English city of York. And he promised that the company would be run by "capitalists from all nations."

Assessing the pamphlet, *The Athenaeum* questioned Lesseps at every juncture. The journal felt that the cost projections were too low, the time estimates for the canal's completion were too optimistic, the technical difficulties too great, and the political issues too intractable.

In the editors' view, the Suez Canal could easily become "a dangerous interference with the existing conditions of intercourse between Europe and the East." The *Edinburgh Review* echoed these concerns and wondered about the difficulties that the Red Sea might present to shipping. This was only the first of many salvos fired by the *Review* against the canal, and over the coming decade, it remained a harsh critic.[3]

Lesseps left England that summer with his rose-tinted glasses intact. He was sanguine about Palmerston's opposition, and he was so convinced of the virtues of the canal that he could not imagine that the majority of the British public would fail to be persuaded. He wrote to the Emperor Napoleon that the English government would carefully consider the merits and that he foresaw no long-term issues. That was, to put it mildly, a positive spin on a less-than-positive reception in London. But in this, as in much else, Ferdinand's belief that what he wanted to happen would happen gave him the confidence to proceed, and, in this as in much else, he was ultimately vindicated.[4]

The next years consisted of a whirl of activity for the nascent company. The engineers Linant and Mougel oversaw a series of commissions and surveys dispatched by both partisans of the canal and opponents, and in each major European capital, advocates of the canal lobbied businessmen and politicians on its behalf.

During this period, however, not a single stone was removed from the isthmus; no final blueprints were drawn up; no financing was arranged. As late as the summer of 1858, in fact, the Suez Canal remained an idea, one that was known throughout the world, but still just an idea. It had its passionate devotees and its implacable enemies, and until the end of the 1850s, it seemed that the enemies had the advantage. In order for Lesseps and Said to succeed, the canal would have to be funded and then built. In order for its adversaries to prevail, the status quo would have to be maintained. That gave the opponents the edge. Keeping things the way they are is always the path of least resistance.

Lesseps himself was hardly sedentary, and he maintained a frenetic schedule. Every great engineering project requires years of planning before it begins. Then, as now, preparatory studies were needed, and the elaborate process of financing the venture had to be determined. The political issues continued to be intractable, even byzantine. Lesseps remained an active correspondent, and his letters and journal entries run to thousands of pages. The diplomatic traffic concerning the canal was heavy, and British, French, and Turkish officials devoted considerable time and effort to the Suez question. Yet, although these years helped transform the dream into an actual undertaking, they can seem static. The politics of the Suez Canal in these years largely consisted of what diplomat X said to diplomat Y and how Lesseps met yet again with banker Z or world leader Q and rehashed the arguments that he had first laid out within months of the original concession.

Between 1855 and 1858, Lesseps spent months in England, months in France, months in Egypt, and months in Constantinople. He passed considerable time in transit between these places, and he also traveled to Trieste and Vienna, Spain and the Netherlands, and even to the Sudan. For all the layers of intrigue, the issues were simple: Lesseps and his allies wanted to build a canal; the Ottomans wanted to avoid offending either the British or the French and to retain whatever tenuous control they had over Egypt; Said wanted a canal, but only if that didn't entail alienating the English or the Ottomans; the British government was opposed; and the French supported the idea, but not at the expense of outright English hostility.

Recognizing these dynamics, Lesseps massaged his allies. In meetings with Said, he reassured the Egyptian leader that the difficulties being encountered would not impair Said's authority or undermine his ambitions. In correspondence with Napoleon and Eugénie, he touted the advancement of the project and stroked the emperor's vanity. In Constantinople, he stoked Ottoman resentment at being seen as a handmaiden of Europe and called on Turkish officials to make their own decisions without bowing to pressure from the British. In France, he played the part of the energetic entrepreneur serving the cause of civilization and tweaking the British lion in the process. And in his visits to England, he lobbied merchants and civic leaders whose fortunes were tied to British exports and to increased world trade.

Each of these interlocutors also pursued a distinct policy in these years. First there was Said, whose ardor began to cool. Bruce, the British consul, continued to pressure the viceroy, and the constant drone of negativity had its effect. Though Said rarely confronted people directly, he made his discomfort apparent by becoming less accessible. Lesseps was no longer granted the same access, and his letters were not always promptly answered. As time passed, Said acted less like a friend and ally and more like an intransigent partner. Business deals are often the death of friendships, and if Lesseps mourned the deterioration of his personal rapport with Said, he gave no sign. For his part, Said was at times warm toward Ferdinand, and at other times distant and mistrustful. He had a rational fear that Suez carried hidden costs for his country and for his own prestige. And as the full dimensions of the undertaking became apparent, Said began to feel that Lesseps was using their friendship to manipulate Egypt and its ruler for his own purposes.

Said honored his commitments, however, and without his generous allowance to Lesseps, little of the needed preparatory work and public-relations campaign could have been funded. In the fall of 1855, Lesseps assembled an international commission of engineers to conduct an independent audit of the feasibility of the plan. Composed of some of the most prestigious men of Europe—including Negrelli of Austria, who had been part of the Study Group in 1847—the commission was instructed by Lesseps to cast a critical eye on the initial surveys of Linant and Mougel. He told them that, though the viceroy preferred the direct route, they were not to be bound by that preference but, instead, to come up with their own best conclusions as to what was practical and preferable.

That December, the commission tested the soil of the isthmus, assessed the problems of constructing the jetties at Port Said, and examined alternate routes. It became clear by early 1856 that the commission was going to issue a favorable report, though the British engineer was less persuaded than his colleagues. He argued for using the Nile to feed the canal rather than the salt water of the two seas; locks could then be installed at either end in order to keep the sea water and the fresh water separate. However, the engineer did not question the basic viability of a Suez Canal that went directly from the north to the south, and the final report of the commission declared, "The execution is easy; success is assured and the results will be immense for the commerce of the world."[5]

These findings were soon disseminated to the general public, and the report became a public-relations weapon in the canal's favor. In addition, on January 5, 1856, from his palace in Alexandria, Said issued a second act of concession, which supplanted the initial document of November 1854. This new concession once again permitted the formation of a company dedicated to the creation of a canal between the Gulf of Pelusium in the Mediterranean and the port of Suez on the

Red Sea. But whereas the first concession had been vague about the route, the second left no doubt that the canal would be a straight line through the isthmus. Said also reaffirmed that a freshwater canal would be dug from the Nile through the Valley of Toumilat to the isthmus at Lake Timsah. And he instructed the company to build ports at both Timsah and Pelusium.

Other articles of the concession specified the generous exemptions that the company would enjoy: no import or export duties on any material relating to the construction; ten years of tax-free use of whatever land it brought under irrigation as part of the works; and the free right to any mines or quarries. Said authorized Lesseps to form a Universal Company of the Maritime Canal of Suez, and promised that the company would have "the hearty co-operation of the Egyptian government" and that all government officials would aid and assist the company in whatever way possible.

There were, however, two sticking points, both of which would later threaten the completion of the canal.

One, the concession stipulated that four-fifths of the workers would be Egyptians. It was understood that the only way Egyptian workers could be provided was through the corveé, which though not exactly slavery, was not free labor either. Several months later, Said issued a supplementary decree concerning "native workmen." They were to be supplied by the Egyptian government and paid by the Canal Company. Men were to be paid between two and three piastres a day, while children under twelve were to be paid one piastre, but given full rations. The company was responsible for supplying drinking water, rations, tents, and transportation costs. Skilled workmen such as masons and carpenters were to receive whatever going rate the Egyptian government paid for such skilled services on other public-works projects. The corvée had been used for work on major Egyptian public-works projects for centuries, and Said's decree seemed innocuous at the time. No one had objected before.

The other complication was an amendment concerning the Ottomans. Though the concession was sufficient for the organization of the company, Said made it clear to Lesseps that work could only begin if the sultan agreed. In a letter addressed to "My devoted friend, of high rank and birth, M. Ferdinand de Lesseps," Said wrote, "As to the works relating to the boring of the Isthmus, the Company can execute them itself so soon as the authorization of the Sublime Porte has been accorded to me." Said had stated his position clearly: the company could organize and prepare, but the canal could only be constructed if the Ottoman sovereign gave his sanction. That should have been the end of the debate, but Lesseps chose to interpret matters differently.

The second act of concession then allowed Lesseps to formalize the statutes of the Universal Company. The company was to be capitalized at two hundred million francs (which was the equivalent of eight million pounds), divided into four hundred thousand shares worth five hundred francs each. Share certificates were to be issued in multiple capitals, written in Turkish, German, English, French, and Italian. The society of shareholders would be administered by a council consisting of thirty-two members who would serve for eight years, and this council would govern the company. There would be a meeting each spring to which all people holding at least twenty-five shares would be invited. The corporate domicile of the company would be in Alexandria, while its administrative offices would be in Paris.[6]

A second act of concession; Said's decrees about native workers; and Lesseps's statutes of the company. These three documents, dry and legal though they were, set the future course of the canal and the company that would build it. Though the corporate structure was to evolve over the coming years, and the nature of the work would as well, the outlines were established in 1856.

The decrees would not have been issued without the work of the international commission, which was the first of several groups that trekked to the Isthmus of Suez to report on the canal and the engineering challenges it presented. In the early stages especially, the backers of the canal needed the seal that the engineering community could give. Without their approbation, Lesseps could not effectively rebut the naysayers. He could not tell skeptics in England that their objections were not supported by the latest science, and he could not convince bankers and potential shareholders that their money would be well spent. He was not an engineer, and, other than Linant and Mougel, none of the initial partisans were either. That was a weakness that opponents tried to exploit. In England, Robert Stephenson claimed that the canal presented far too many technical problems, while in France, the Saint-Simonians challenged Lesseps's credentials and competence.

Enfantin should have realized that events had passed him by, but he was not one to fade gracefully. If anything, signs that he was being eclipsed led him to intensify his campaign in the fall of 1855. One of Napoleon's ministers, the Comte de Persigny, expressed misgivings about the canal and sounded out various influential people in England to scuttle the plan. Allied with Persigny, Enfantin and his followers waged a smear campaign against Linant and Mougel. Lesseps was weakest at the level of his technical expertise, and until that first international commission reported in early 1856, the blueprint for the canal derived primarily from the work of Linant and Mougel. Undermine them, and Lesseps would be crippled.

Enfantin also kept lobbying the imperial court. He reminded Napoleon that the Saint-Simonians had been promoting engineering projects for two decades, whereas Lesseps was a disgraced diplomat who had never overseen anything larger than a country estate. Enfantin also argued that a canal would be in the imperial interests of France and would allow France to counteract the expansion of Russia into the Near East, just as the Crimean War had prevented Russia from expanding into the eastern Mediterranean. But leaving the canal to a naïf like Lesseps and his motley band of Franco-Egyptian civil servants would ensure the premature death of a brilliant idea.

Enfantin then enlisted the support of Robert Stephenson, who was both a member of Parliament and a revered engineer. In addition to his standing in politics and business, Stephenson wielded considerable influence as a leading member of several professional organizations. In the mid-nineteenth century, the last medieval guilds finally disintegrated and gave way to the professional "institutes," such as the Royal Geographic Society and the Institution of Civil Engineers. Found throughout Europe, these societies were loosely organized, but they served the function of setting standards of what was and was not acceptable in the various professions. They had the de-facto capacity to terminate a nascent project with a negative report, or to guarantee its success with a positive endorsement. Stephenson was respected, and his vote could doom a speculative new endeavor.

While searching for allies, Enfantin alternated his letter-writing to the emperor with missives to his acolytes. With the emperor, he was formal and deferential; to his followers, he could be crude and haughty. In one letter, he went on at length about his realization that Lesseps's name rhymed with a common mushroom—*"le ceps."* He also vented his eschatological paranoia. He said that Lesseps was "the herald who cries from afar and makes the nations take note of our marvelous promise. He makes a new plan, but so what? What does a plan matter? What matters is the dream of building a sainted temple, the temple of the future, not of the past, the Temple of Peace. . . . We are certain to have a share of glory, in heaven and on earth in the present century and in future centuries, as we now have a share of misery."

But even in his more politic exchanges, his hatred of Lesseps was apparent. After several such letters, the emperor's chamberlain finally responded in December 1855, saying that the emperor had been informed of Enfantin's objections, "but His Majesty instructed me to tell you that he does not believe that he ought to intervene in this question." With that curt note, Enfantin was dismissed, never to be taken seriously again in the debate over the canal. The emperor had not yet taken a public stand in favor of the Suez Canal; that would come later. But in his response to Enfantin, as well as to others, Napoleon sent subtle signals of his approval.[7]

Stephenson, however, was just getting started. He did all that he could to impugn the credibility of the project and the motives of its backers. Throughout 1856 and 1857, he emerged as the expert of choice in arguments against the canal. Though there were no "pundits" in mid-Victorian England, Stephenson played the functional equivalent. For those who wished to refute a pamphlet published by those in favor of Suez, invoking Stephenson's name was an effective tactic. During heated debates in Parliament in 1856 and 1857, he rose to declaim against the canal, and his words were widely discussed.

Like Palmerston, Stephenson was long on bluster; unlike him, he was short on tact. In the summer of 1857, both went too far. Over the previous year and a half, Lesseps had been campaigning for the canal in England, either in person or through proxies. In person, he would speak to the Chamber of Commerce in a city like Liverpool or Birmingham, often after a dinner and lavish reception replete with ceremonial toasts. The speech he gave varied little, and touched on the economic benefits for world trade and for the British Empire above all. He was an adroit propagandist, but in his passionate defense of the canal, he was utterly sincere. Audiences intuitively knew that he was speaking from the heart, and even if they questioned his logic, no one doubted his integrity.

No one, that is, until Palmerston and Stephenson spoke in Parliament in the summer of 1857. Lesseps had been in Britain since April, and this three-month sojourn was his most extensive period in the country. He appeared at dozens of public meetings, and hundreds of thousands of people either met him, heard him, or read about him. Palmerston was unmoved. He remained distrustful of France, and was still unwilling to grant that the canal was technically possible. Lesseps concluded that Palmerston was unlikely to change his position, ever. Rather than fight the prime minister directly, Lesseps had redoubled his efforts to win over the public.

Slowly, British opinion began to shift. Several powerful voices defected from Palmerston and spoke in favor of the canal, including Gladstone, who agreed with the basic contention that the canal would be good for free trade, good for the British, and good for civilization. Each week, Lesseps accumulated more endorsements from chambers of commerce. After a meeting in Newcastle-upon-Tyne, the chamber council voted in favor of the Suez Canal and declared "that it would be the most advantageous to the world, and to this country most especially, if the present long and tedious route for shipping between Europe and India could be superseded."[8]

But this cascade of support only intensified the determination of Palmerston and Stephenson. Newcastle may have tipped the scales; Stephenson had been born near the city, had built its most prominent bridge, and may well have felt betrayed by the actions of its Chamber of Commerce. For his part, Palmerston had been re-elected by a considerable margin that spring. He believed that he understood British interests better than the democratic rabble, and he was secure enough politically to flout public opinion on this issue. Lesseps's successful tour of the country convinced the two men that a more aggressive response was necessary, and that summer they mounted their most intense assault. It was Stephenson, however, who bore the brunt of Lesseps's infuriated response.

Palmerston received Lesseps one more time, and greeted him by saying, "Well! You're going to make war with us? You've stirred up the passions of England, Ireland, and Scotland about this Suez Canal question." The two then rehearsed what had by then become a familiar and tedious discussion, but with a new twist: Lesseps told the prime minister that the public was now in favor of the canal. Palmerston realized that Lesseps was right.

On July 7, 1857, there was a heated debate in the House of Commons on the Suez Canal question. The defenders of the canal were putting pressure on Palmerston to cease his obstruction, and they had called on the government to change its stance toward the sultan. It was common knowledge that Lord Stratford was holding the canal hostage through his private démarches to the Sublime Porte, insinuating that if the sultan affirmed the canal concession there would be a rupture between Great Britain and the Ottoman Empire. That policy looked less wise in light of the grave crisis that the British were facing in India. The Sepoy Rebellion had erupted in the spring, and it was unclear whether there were sufficient numbers of troops in northern India to suppress the revolt. Though the British brutally reasserted control, Palmerston's arguments against Lesseps suddenly seemed weaker. The Suez Canal would make it possible to send reinforcements to India in half the time, should another disturbance occur. As a result, some who had been ambivalent about the canal now called for its construction.

Palmerston rose and offered a pallid defense. "Her Majesty's Government," he began, "certainly cannot undertake to use their influence with the Sultan to induce him to give permission for the construction of this canal, because for the last fifteen years Her Majesty's Government have used all the influence they possess at Constantinople and in Egypt to prevent that scheme from being carried into execution. It is an undertaking which, I believe, in point of commercial character, may be deemed to rank among the many bubble schemes that from time to time have been palmed upon gullible capitalists." He went on to deride the practicality of the canal, except at immense and unprofitable expense, and to warn that, even if it could be done, it would lead to a shift in power between Egypt and the Ottomans that would hurt Britain. He also suggested that, while persuasive, Lesseps had ulterior motives, and that "the object which he and some of the promoters have in view will be accomplished even if the whole undertaking should not be carried into execution." In short, he accused Lesseps of promoting a scheme for profit and forcing the Egyptian government to underwrite the canal company regardless of whether the canal itself was ever constructed.

This speech brought an immediate response from Lesseps, who sent a circular to more than a hundred trade associations, newspapers, journals, and chambers of commerce with a point-by-point rebuttal of Palmerston's arguments. He called on "the commercial classes of England" to decide whether their prime minister's stance was in their interest, and he angrily denied that he was motivated by any base interest in profit, or that he had any "designs upon the pockets" of English capitalists.

On July 17, the issue was debated again in the Commons. Palmerston repeated his objections, and once again termed the project "one of those bubble schemes which are often set on foot to induce English capitalists to embark their money upon enterprises which, in the end, will only leave them poorer, whomever else they make richer." He then turned to Stephenson for support. Stephenson rose and said that "he would not venture to enter upon the political bearings of the subject . . . but would confine himself merely to the engineering capabilities of the scheme." He explained that he had surveyed the isthmus many times, and that the canal presented too many engineering obstacles at too great a cost to make it a viable commercial alternative to the rail link.

Though Stephenson had only seconded what Palmerston said, he bore the brunt of Lesseps's wrath. As a private French citizen, Lesseps could not credibly demand an apology from the prime minister of Great Britain. But he could go after Stephenson. Arguments against the merits of the canal Lesseps could accept, but assaults on his personal honor he would not. Hearing of what had transpired in Parliament, Lesseps fired off an angry letter to Stephenson demanding clarification, either in writing or in person, accompanied "by two of your friends." In the idiom of the day, this was an ultimatum. Stephenson could either apologize for his scurrilous remarks or meet Lesseps for a duel, accompanied by seconds, to settle the affair permanently.

For all his bluster, Stephenson had little interest in pistols at dawn. No one dueled in England anymore, and in any event, Stephenson was more committed to preventing the canal from being built than to dismantling Lesseps's reputation. The next day, he sent a conciliatory reply. "Nothing could be further from my intention," he said, "in speaking of the Suez Canal the other night in the House of Commons, than to make a single remark that could be construed as having any personal allusion to yourself. . . . When I said that I concurred with Lord Palmerston's opinion, I referred to his statement, that money might overcome almost any physical difficulties, however great, and that the undertaking if ever finished, would not be commercially advantageous." Lesseps pronounced himself satisfied as far as his honor was concerned but still perplexed that Stephenson would take it on himself to disagree with the bulk of the international engineering community that the canal was, in fact, quite feasible.[9]

This was the most dramatic confrontation between Lesseps and his adversaries in England. Though British antagonism did not abate over the next ten years, all sides understood that the balance shifted in Lesseps's favor after the summer of 1857. The backers of the canal would have to contend with the dangerous hostility of the British government for years to come, but never again would Lesseps need to attend so assiduously to converting the British public.

Before this denouement, however, the stance of Her Majesty's Government had forced Lesseps to spend considerable time in Constantinople, and to pay more attention to his relations with Said. In 1856, Lesseps passed another few weeks in the Ottoman capital, attempting to counteract the corrosive influence of Lord Stratford. That was not an easy task. Stratford had kept up a constant drumbeat that the canal would allow Egypt to declare its independence from the empire, and that argument was repeated to other Ottoman officials by other British representatives in other cities.

In Paris, as the treaty ending the Crimean War was being negotiated, the British ambassador, Lord Cowley, met with the Ottoman prime minister, Ali Pasha, in late March 1856. Cowley warned that the real aim of Said was to secure his independence from the Porte. Ali said that he agreed but could find no adequate reason to prevent preparatory work from going ahead. Said could not be refused without good cause, at least not without pushing him further into the arms of Lesseps and France. Instead, Ali told Cowley that he planned to find a way to make the canal so onerous for Said that the viceroy would drop it voluntarily. The idea was to tie the sultan's approval to conditions that Said would find unacceptable, such as the demand that any fortifications along the canal be manned by Turkish troops dispatched by the sultan. Said would find this an untenable infringement on his autonomy. He would refuse to honor that condition. The sultan would then refuse to permit the construction, and Said would then have no choice but to withdraw his support from Lesseps.[10]

Ali Pasha's elaborate trap was clever, but it was never laid. It was also an indication of just how few tools of coercion the Ottomans possessed. On paper, the empire still looked vast, encompassing more territory than most of its European rivals. Looks were deceptive. The sultan's control over his realm decreased with each passing mile from

Left: Ferdinand de
Lesseps (1805–1894)
*(Courtesy of Getty Images)*

Below: Napoleon and
his army in Egypt, by
Jean-Léon Gérôme
*(© Christie's Images/CORBIS)*

Left: Muhammad Ali Pasha, father of m[...]
Egypt (© *Bettman/CORBIS*)

Right: Père Enfantin in his usual
eccentric clothes (*Granger Collection*)

Below: A group in the slave market of
Cairo, by David Roberts (*Author's collection*)

*The Death of Sardanapalus*, by Eugène Delacroix (© *Philadelphia Museum of Art/CORBIS*)

Muhammad Said Pasha, ruler of Egypt (1854–1863) (*Reprinted by permission of the Association de Souvenir de Ferdinand de Lesseps et u Canal de Suez*)

Lord Palmerston, prime minister of Great Britain (1855–58, 1859–65) (© *Hulton-Deutsch Collection/CORBIS*)

Emperor Napoleon III of France and his wife, Empress Eugénie (*The Illustrated London News Picture Library*)

Caricature of Ferdinand de Lesseps, se[...] the two continents (*Reprinted by permission [...] sociation de Souvenir de Ferdinand de Lesseps et [...] de Suez*)

Lesseps and the leading engineers of the Suez Canal (© *Bettman/CORBIS*)

Right: Ismail Pasha, ruler of
Egypt (1863–1879) *(Reprinted by
permission of the Association de
Souvenir de Ferdinand de Lesseps et
du Canal de Suez)*

..mail's palace on the shores of
..sah, next to the Suez Canal *(Re-
..ermission of the Association de Souvenir
..d de Lesseps et du Canal de Suez)*

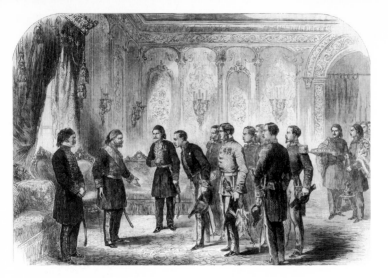

the consuls general o
occasion of his acces
the pashalic of Egypt
*Illustrated London News*
*Library)*

Right: One of the
dredges designed
specifically for work
on the canal
*(Reprinted by permission*
*of the Association de*
*Souvenir de Ferdinand*
*de Lesseps et du Canal*
*de Suez)*

Left: T
of Ferd
Lessep
Ismaili
*(Courtes*
*Lesseps f*

The Suez Canal works: excavation at El-Guisr *(The Illustrated London News Picture Library)*

The opening of the Suez Canal: blessing the canal at Port Said in the presence of the imperial and royal visitors *(The Illustrated London News Picture Library)*

The opening of the Suez Canal: the procession of ships in the canal
(*The Illustrated London News Picture Library*)

The city of Port Said in the twentieth century (© *Hulton-Deutsch Collection/CORBIS*)

Constantinople, and though he still received tax revenue from far-flung provinces, the cohesion of the empire depended on a tacit agreement between the capital and the provinces: the sultan would not infringe on the autonomy of provincial governors and they would not challenge his authority.

In the meantime, the imperial city itself was changing. Constantinople in the mid-1850s reflected the cultural schizophrenia of the Ottoman Empire. The metropolis was a patchwork of modern and medieval. There was an old section that had hardly changed since the Ottoman golden age, and there was a new city modeled after Europe. The old city spread out below the Hagia Sophia, which had been built by the Byzantine emperor Justinian, and the Blue Mosque, which had been constructed by a seventeenth-century sultan to rival it. That city was a maze of narrow lanes and cramped quarters. There was a street of tinsmiths, a street of butchers, and a street of tailors—just as there were centuries before. The new sections across the Golden Horn were laid out in a grid, with regular street-cleaning, lighting, and a municipal administration. The sultan built a European palace in the new city, but his wives were still sequestered in a harem guarded by eunuchs. High officials often wore Western clothes, but sometimes they donned the traditional robes of the Ottoman aristocracy. Yet, though the city presented a chaotic juxtaposition of modern and traditional, coffeehouses were everywhere, offering an escape and a respite. And in the morning, fishing vessels still brought in their catch from the Bosporus, as they had for millennia, and in the evening, the rich and powerful, as they had since the time of Constantine the Great, still opened their windows so that the breezes coming off the water would cool their mansions.[11]

That was the city that welcomed Lesseps at least once a year, starting in 1855. He returned early in 1856, just before the sultan issued a vital imperial decree committing the Ottoman Empire to a path of reform and to harmonious relations with the European powers. Yet, though the empire was in flux, the stance of Ottoman officials toward the canal was static. Each new visit brought the same results. Each time, Lesseps made the rounds from the Sublime Porte to the various European embassies. Each time, his arrival triggered a flurry of diplomatic exchanges, and by the time he left, the situation would remain unchanged. The British tried to keep the Ottomans from succumbing to French entreaties; the French tried to prevent the British from swaying the sultan; and the sultan managed to satisfy both by doing nothing.

As of the late spring of 1856, therefore, Lesseps still did not have Ottoman approval. That might have weighed on someone else, but he took comfort from the fact that he had not acquired any new liabilities. Personally, he didn't think that the sultan's formal approval was necessary. Said, however, was growing uneasy in its absence, and a vacillating Said was something Lesseps could not afford. For long stretches in 1856, Lesseps was absent from Egypt, whether he was in Constantinople, sipping coffee with the vizier; in Vienna, making his case to Prince Metternich; in Paris to meet with the cream of Europe's diplomats, who were assembled to negotiate the end of the Crimean War; or in England, fighting with Palmerston. All the while, however, Lesseps stayed in touch with the Egyptian ruler, either through direct correspondence or through representatives in Cairo and Alexandria. He was kept informed of the viceroy's mood, and by the year's end, it was clear that Said's support was weakening.

His doubts notwithstanding, Said honored his financial commitments. He had promised his support, and he delivered. He paid for surveys of the isthmus as well as for the publication of *L'Isthme de Suez: Journal de l'union des deux mers* (*Journal of the Union of the Two Seas*). At first, the journal was simply an expensive, effective piece of propaganda, but it soon evolved into a professional newspaper that supplied consistent, albeit biased, information about Egypt and the canal to the French public. It was partisan, but no more so than most papers of the time in Paris or London. Journals were expected to take sides, and to be organs for one political party or another. While *L'Isthme de Suez* was limited in scope, it continually and effectively presented the canal as a noble work blessed by talented engineers and opposed by pernicious forces.

Said's support, while costly, did not reflect his shifting mood. Granted, it is difficult to know what his mood was. Like his father and like his successors, he left no private papers and no personal record of his approach toward governing. He is known almost entirely through the eyes of others, and the portrait of him derived from British sources is not flattering. British officials in Egypt portrayed the viceroy as a weak-willed dimwit, but their record of him is suspect. Having endorsed the canal within months of assuming power, Said was at odds with the British for his entire reign, and British animosity colored their impressions of him. Lesseps had a more favorable opinion, but neither he in his letters and diaries nor French officials in their memoranda were prone to editorializing about Said's character.

Reading between the lines of the multiple and incomplete set of letters and dispatches, it seems that Said was more aware of what he was doing than his poor reputation would suggest. While the British depicted him as inept and overmatched, he was hardly passive. He personally went to Constantinople in the fall of 1856 to gain the sultan's favor, though he came away empty-handed. With or without the sultan, however, he had the power to scuttle the canal. Lesseps understood that. During the winter of 1856–57, Said left for a long journey to the Sudan. Egypt had for centuries claimed control of the northern portion of that vast region to its south, and the viceroy treated the trip as an extended survey of his realm. Lesseps took the opportunity to spend an uninterrupted period of time with Said, hoping perhaps to mimic that glorious week in November 1854. Having enchanted Said once in the desert, Lesseps sought to inoculate the monarch permanently against the canal's opponents.

This time, Lesseps was not successful, and Said did not embrace Lesseps's rosy words. The idea of the canal was causing tension between Said and Britain and between Said and the sultan, and those relationships were more important to him than a not-yet-constructed trench through the desert.

Lesseps did not linger in Egypt after the Sudan expedition. In fact, Said's anxiety about Britain was one reason for Lesseps's publicity tour of England, Ireland, and Scotland that spring, which in turn led to the imbroglio between Lesseps, Stephenson, and Palmerston. By the fall of 1857, Lesseps was clear about what he had to do next. Though the diplomatic situation was unresolved, the wavering of Said and the animosity of Britain could only be countered by pushing forward. Further delay would be tantamount to defeat. Little movement had occurred after three years of effort, and Lesseps decided to force the issue. He would get the support of the Emperor Napoleon, found the company, sell shares, and begin the work, whether or not Said, the sultan, or Palmerston agreed. It was a risky move. It would take a year, but it would succeed.

# CHAPTER TEN

# THE EMPEROR AND THE ENTREPRENEUR

THESE YEARS OF constant motion with little to show for it might have dented the confidence of a man more prone to self-doubt or introspection, but Lesseps never flagged. He had set a goal, and until every avenue had been exhausted, he would keep going. Tenacity alone would not have been enough, however. In fact, all the organizational skills and vision in the world might not have been able to combat the hostility of the British, the passive aggressiveness of the Ottomans, and the waxing and waning of Said's passion. But Lesseps had one other advantage, one that everyone was aware of but to which he did not draw attention. He had good connections in the diplomatic world, yes, but he was also the cousin of the empress of France. Most people knew that, but few knew quite what it meant.

In the 1860s, Lesseps and Eugénie de Montijo became quite close, but before that, their relationship was something of a mystery. He had watched her grow up during his years in Spain, and she wrote him repeatedly during her quick courtship with Louis-Napoleon to seek his advice. After that, they maintained a correspondence, but the few surviving letters do not suggest more than a polite, familial fondness between them.

How close they actually were is less important to the story of Suez than how close people perceived them to be. It would have been gauche for Lesseps to make explicit mention of his relationship to Eugénie, but in truth he did not need to. It was common knowledge. The extent of his influence with her, and, by extension, with the emperor himself, was not. The uncertainty was a boon for Lesseps. Though no one knew whether Eugénie interceded with the emperor on Ferdinand's behalf, prudence dictated giving Lesseps a wide berth.

Uncertain how much hidden power Lesseps possessed by virtue of his familial proximity to the throne, most erred on the side of assuming that he had too much, rather than gambling that he had very little. If they were wrong, at least they would not alienate Napoleon III; and if they were right, they would please the emperor by pleasing Lesseps.

The vague nature of his influence at court also boosted the prospects of the canal away from Paris. Farther removed from the circles of European intrigue, Egyptian elites and Ottoman grandees heard rumors that Lesseps was an emissary of the emperor. They assumed that Lesseps's ambition was a direct extension of Napoleon's goals in the Near East. That Lesseps had no official position was taken as a sign that the emperor wanted the canal to be built. Trained in the ancient arts of obfuscation, the aristocrats of Cairo and Constantinople thought they were seeing through a clever ruse. They assumed that, by dispatching a personal emissary who could not be directly linked to the crown, Napoleon was signaling that the canal was dear to his heart and was so important that he did not want to risk sending a high-ranking official who might be publicly rebuffed. Instead, he sent Lesseps. Many officials assumed that a direct snub of Lesseps was the equivalent of slapping the ruler of France in the face, and though they did not always do what Ferdinand asked, neither did they say no.

In 1856, Napoleon III and his empress were in their prime. The Second Empire had become a glittering state, and that spring, Paris welcomed the diplomats of Europe. The primary reason was to settle the morass of the Crimean War, which France, Britain, and the Ottomans had won, but which had hardly left Russia crippled. The gathering of so many diplomats ensured that a whole range of subjects would be canvassed, including the question of Suez. Not high on any country's agenda, the matter of the canal was relegated to casual discussion. It was talked about not in formal sessions but over dinner or drinks at the emperor's palace. Though the dispute was not resolved, it had become a subject for the powers to consider, and that was a victory for Lesseps.

While the conferees restored the balance of power, Napoleon took the opportunity to show off his new city. Paris was in a flurry of construction. Under the direction of Baron Haussmann, wide avenues were being carved through the old city. Thousands of the poor were displaced, and whole districts were razed to the ground. Hundreds of millions of francs were spent. And the Paris of the Rue de Rivoli, the Boulevard Saint-Germain, the Opéra; the Paris of grand railroad stations and equally grand cemeteries; the city of bourgeois *boulangeries* and uniform building codes, of stately suburbs and parks to stroll through on a Sunday afternoon, began to take shape. Paris had been a filthy place, with sewage emptying into the Seine next to pipes that carried drinking water in the other direction. Cholera was endemic, but the inhabitants of the city had accepted the risk of disease and the chaos as acceptable costs for living in the hub of power and culture, and they disapproved of Haussmann. Yet, though they did not embrace his reforms, they were not allowed to prevent his renovations.

Haussmann conceived of the master plan, but he was only a tool of the emperor. Napoleon III, born Charles Louis Napoleon Bonaparte in 1808, nephew of the Corsican who nearly ruled the world, was an unlikely monarch, except for his illustrious name. Before he turned forty, he had spent most of his life either in exile in Austria, Italy, England, and New York, or in prison for the pathetic coup attempt he staged in 1840 along with his ragtag followers, who thought that the French army would rally to his side at the very mention of the name Bonaparte. After escaping from prison in 1846, he carried on in foreign capitals and developed a reputation for enjoying the nocturnal company of a wide variety of women, many of whom were explicitly seeking a brief encounter with a Bonaparte.

Were it not for the chaos of 1848, Louis-Napoleon would probably have lived the louche existence of a man without a country, seeking a role that was worthy of his dreams but content to play cards, get drunk, and wake up with a new stranger each morning. Europe of the nineteenth century was filled with exiled princes, and though they were figures of pathos, it wasn't a bad life. Napoleon, however, was enamored of his illustrious name, and, for all of his wastrel appetites, he kept his focus on France. Surrounded by a coterie of loyal retainers, he saw 1848 as an opportunity to revive the Napoleonic ideal.

In a swirl of moves and countermoves, Louis-Napoleon was elected as a deputy in the republican Assembly, and then, stunningly, won a national election in December 1848 as president of the Second Republic. That republic was short-lived, but Louis-Napoleon was only getting started. After three years as prince-president, he transformed his regime into a popular autocracy and staged a coup from above in December 1851 as thirty thousand loyal troops occupied Paris and arrested thousands of opponents. Freedom of the press was drastically curtailed, and political opposition was silenced. He called for a national referendum to ratify his authority, and won more than seven million out of eight million votes cast. Less than a year later, he called for a national referendum to proclaim him not president but emperor. "The Empire is peace," he announced, "because France desires it, and when France is satisfied, the world is peaceful." The vote was even more lopsided than it had been a year before. Though three hundred thousand voters disagreed, 7.8 million others voted in favor of making Louis-Napoleon into Emperor Napoleon III, and allowing his heirs to inherit his throne. With that, the Second Republic came to an end, and the Second Empire began.

Louis-Napoleon had power, but he had no coherent program. "The name of Napoleon is a program in itself," he had said in 1849, and for the eighteen years of the Second Empire, his name was the only constant. As one later writer quipped, "He had been a man of one idea; and when it was accomplished, he was left without one." It is true that the emperor did not have a detailed blueprint for France's future. Having spent his life trying to restore the Bonaparte name, he had not wasted much time thinking about what he might do in the unlikely event that he succeeded. Many leaders have a short horizon and find themselves improvising as they go along. Napoleon lacked a comprehensive plan for what he would do with his power, and, to his lasting discredit, he was cursed with a chattering class that despised him for his limitations.

Napoleon may have ruled France, but the scorn of the intelligentsia has forever tarnished him. Though much of that tarnish may be merited, it has been difficult to disentangle Napoleon III as portrayed by the likes of Victor Hugo and Émile Zola from the Napoleon III who was elected by an overwhelming majority of French citizens, and who commanded their loyalty for a considerable portion of twenty years.

Zola was blunt in his disdain. Napoleon had "more imagination and more dreams than judgment." He had dreamed of reviving the legacy of his uncle, but for a man of Louis-Napoleon's limitations, that legacy was all form. It meant an empire and titles and palaces and grand projects and an extension of French power abroad. "What an idiot!" Zola concluded. Others were equally harsh, including the towering figure of Adolphe Thiers. "He's a cretin," remarked Thiers of the prince-president who would soon become emperor. But perhaps the most acute among his critics was the writer and sometime politician Victor Hugo, who wrote of Louis-Napoleon, "France's first mistake was to take him for an idiot; her second was to take him for a genius."

Harsh words, and with no more consequence than lilliputian arrows, at least not at the time. But the subsequent history of France, and the tumultuous collapse of the Second Empire in 1870 followed by the violence of the Paris Commune and the flux of the Third Republic, seemed to vindicate the picture of Napoleon III as a bumbling, fatuous fool play-acting at being emperor. Saturated with such negative images, the Second Empire has been difficult for later generations to approach on its own terms, but in the 1850s, Napoleon seemed on the verge of creating a new dynasty. Whatever people thought of him privately, little of consequence happened in France or Europe during these years without his input.

His stance, however, was often hard to gauge. He was notoriously averse to confrontation. Petitioners were routinely misled into thinking that he had agreed to grant them things that they were subsequently informed he had not. He was reticent in his public statements, and somewhat awkward physically, though he was a superb horseman. According to those close to him, however, his reserve in public did not do justice to his surprisingly active mind.

For much of his life, he clung to the notion that the name Bonaparte would overturn injustice. Judging from his prison writings, he believed that he had been placed on earth for a grand purpose. Detained for six years in the château of Ham, in the northeast of France, he read widely; there was, after all, not much else to do. He studied government and philosophy, and he was particularly entranced by the writing of Henri de Saint-Simon. He fell in love with Saint-Simonian ideas about destiny, progress, and industry, and became ever more convinced that it was his fate to enhance all three. He embraced the notion that history is made by great men: "I believe that there are certain men who are born to serve as a means for the march of the human race. . . . I consider myself to be one of these."

Once in power, Napoleon III quickly found an eligible wife in Eugénie de Montijo, and they spent an active few years trying to have a child. That took somewhat longer than planned, but in those barren three years between 1853 and 1856, Napoleon sent French armies to the Crimea and embarked on a successful state trip to visit Queen Victoria. While in England, he was given a tour of the Crystal Palace exhibition in London, which touted the technical marvels of the time, and he was seized with a vision of an even grander exhibition in Paris. Thereafter, his regime was marked by a series of extravaganzas, both at his courts in Fontainebleau and the Tuileries, and in public spectacles, the most glorious and decadent coming in 1867, just in time to celebrate the wonders of Egypt and the Suez Canal.

But in 1856, that lay in the unknown future. As the delegates came to make peace in Paris, Napoleon celebrated not just his new prominence as the glittering king of France, but the birth of a son who he had every reason to believe would succeed him. The France of Napoleon III became known as a carnival empire, which rivaled even the Versailles of Louis XIV. In spite of the carping of the intelligentsia, Napoleon was popular at home. France, said Napoleon's half-brother Morny, was "so tired of revolutions that all it wants today is a good despotism." Self-serving though that observation was, it also seemed true. Many welcomed the stability and order of the Second Empire, even if this came at the cost of curtailed liberties. And Napoleon, oddly enough, spoke with great admiration for those liberties, even as his extensive police force made sure that no one used them against his regime.

Following the muse of his uncle, Napoleon aspired to make France the pre-eminent power in the world. Crimea was one example; the aggressive colonization of Algeria was another. He also supported moves into Syria, Senegal, Mexico, and Indochina, and though he kept his European ambitions in check, he did go to war with Austria and succeeded, in that limited conflict, in aiding the creation of a new Italian state. It was in this spirit of magnifying the role of France in international affairs that Napoleon slowly became a *canaliste*.[1]

Napoleon was an ardent devotee of industrialization. His reign coincided with the greatest period of railroad construction in French history. On his watch, France began to emulate the English model of urban factories and cities with industrial suburbs. Although the French remained fond of small industry and local work, the changes were substantial. French society might have been transformed regardless of who held power. Railroads were being built throughout Western Europe in these decades, by entrepreneurs like Stephenson and Enfantin, and they neither looked to nor asked the political class for permission. Still, for industrialization to proceed, a friendly government was a benefit, and a hostile government was a liability. Napoleon was a friend to industry, and a believer in it.

These passions, for progress and for the glory of his name and the name of France, inexorably led Napoleon to support the creation of the Suez Canal. As his prison writings reflected, his enthusiasm for grand engineering projects was no secret. Though he refrained from aggressively endorsing the Suez undertaking in the mid-1850s, rumor was that he favored it. The emperor took no official position, but there were signs; and in a world of imperial-court intrigue, these signs were widely read.

The emperor's rebuff of Enfantin, for instance, was a clear indication that if the court were to lend its approval it would be to a canal overseen by Lesseps. Then there was the seemingly casual conversation the emperor had with the Turkish foreign minister, Ali Pasha, in April 1856, after a banquet during the Crimea peace conference. Napoleon told Ali that he thought that Suez would benefit everyone. He said that he had been studying the matter and that, though he had no plan to force the issue with England, he couldn't imagine why any reasonable person would wish to prevent its eventual execution. That, in turn, echoed what the president of the French Senate had said several weeks earlier, when delivering a benediction celebrating the birth of the young prince. In front of the emperor and his empress, the senator announced that the child's birth would occasion many blessings, one of which was a joining of the oceans: "The East and the West, which have been seeking since the Crusades but are only now finding each other once more, will marry the two seas and their coasts to release a beneficent flow of ideas, of wealth and of civilization." Saint-Simon could not have put it more eloquently had he been alive to deliver the benediction himself.[2]

In addition, the protector of the Canal Company was Napoleon's cousin Prince Napoleon, known as Plon-Plon, who was the son of one of Napoleon Bonaparte's brothers. The prince had been Napoleon III's official successor until Eugénie had a son. Plon-Plon was widely disliked as a physically repulsive, petulant, ineffectual, and corrupt presence in the emperor's inner circle. He made a fortune trading on his connections to the throne, and he used the proceeds to live grandly in a mansion off the Champs-Élysées. He was corrupt, but he was also savvy. Early on, Lesseps approached him and asked that he be the official patron of the canal project in France, and the prince, never one to pass up a golden opportunity, accepted. For minimal effort, Plon-Plon was guaranteed maximum financial reward if the venture succeeded.

In turn, Lesseps was able to solidify the widespread belief that the emperor himself was a closet *canaliste*. The presence of Prince Napoleon's name on all official correspondence for the project deepened suspicions that it was sanctioned not just by the prince but also by his cousin the emperor.

And, finally, there was Lesseps himself, who continually invoked the emperor's name. Each time he met resistance in London or Constantinople or Cairo, he insinuated that the emperor agreed with whatever course of action Lesseps was suggesting. This use of Napoleon's name infuriated not just English diplomats, but some of the French as well, who felt that Lesseps was circumventing normal channels of diplomacy. In December 1856, Lord Cowley complained directly to Napoleon that Lesseps was using the emperor's name inappropriately and giving the impression that the promotion of the canal scheme was the official policy of the French government. Though Napoleon said that Lesseps did not speak for him, Cowley was left with the distinct impression that the emperor was not displeased. He was correct. Lesseps continued to foster the impression that Napoleon wanted the canal to be built, and Napoleon did nothing to suggest otherwise.[3]

But translating Napoleon's passive support into an active asset was a different matter. By the end of 1857, Lesseps had accumulated all the good will he could use. The canal had been publicized; plans had been debated; assurances had been given; and there were sufficient numbers of ardent people in each of the relevant countries. But two things were missing: money and actual work.

Said was paying for preliminary studies, for the publication of the company newspaper, and for the basic administrative costs. The expense was not inconsiderable, but it was a fraction of the money needed to build the actual canal. Lesseps intended all along for the project to be funded by a joint stock company, and as he learned more about the vagaries of high finance, he became more determined than ever that the company should consist of thousands of shareholders rather than a limited number of wealthy patrons and banks. But, as much as he liked the idea of floating shares to any and all who wished to buy them, he did investigate other financing options.

A number of the large European bankers—the Barings, the Foulds, the Rothschilds—expressed an interest in arranging financing, and Lesseps met with them. His session with Baron James Rothschild in Paris in 1856 may have tipped the balance once and for all. The two men had known each other for some years, and the baron was eager to participate. Rothschild thought the preliminary figure of two hundred million francs sounded reasonable, and he told Lesseps that he would be happy to set the process in motion. The Rothschilds had family members and branches in every major capital, and they could draw on a network of contacts among the political and financial elites of Europe. The baron agreed that the project would benefit mankind and said that his family would be honored to play a part.

Lesseps was delighted, and expressed his thanks. He had one question: how much would Rothschild's services cost? "Mon Dieu!" replied the baron. "It's clear that you are not a man of business. The usual five percent." "Five percent?" Lesseps answered aghast. "On 200 million? But that's 10 million francs!" Lesseps indignantly refused, and said that he would rent offices and arrange the financing himself. Rothschild was not pleased, but decades later his family managed to profit from Suez all the same, when the London branch organized the purchase of millions of canal shares by the British government of Lord Disraeli.[4]

Refusing to fund the venture via the large banks was not, however, a solution to the problem. In order to float shares, the company had to have credibility. Few people would risk the five-hundred-franc price for one share of a company that seemed to have a high chance of failure. Until there was a consensus that the canal was technically feasible and financially practical, it would be impossible to convince large numbers of people to invest. Though Lesseps and his partisans were ready to issue shares as early as 1856, had they tried to, the endeavor would have been a comic failure. Suez did not have the explicit backing of the ruler of France, and the canal had not yet won the ringing endorsement of the engineering community. And however much they celebrated progress, the conservative and cautious middle class in France and England were not yet prepared to entrust the company with their money. As a result, it wasn't until 1858 that Lesseps could seriously consider floating shares. Only after years of speaking in front of hundreds of audiences and having the issue aired in numerous newspapers and professional journals, and only after attending to tedious diplomatic niceties, did Lesseps arrive at the point where the company could be capitalized.

The year 1858 also marked a shift in Napoleon's relationship with Great Britain. On the evening of January 14, Napoleon was on his way to the opera. He was accompanied by Eugénie and his usual coterie of courtiers and guards. Shortly before halting in front of the building, the carriage was rocked by explosions. Three bombs were hurled, one after another, toward the emperor's party. The crowds that had assembled at the entrance scattered in panic, but not before twelve people had been killed and more than a hundred wounded by flying shards of glass and debris. One of the would-be assassins stood at the carriage door, ready to attack the first who exited. The police intercepted him before he had a chance. The empress, unscathed but covered in the blood of one of the injured guards, refused the hand of the theater manager and declared that she would walk alone. The emperor followed her, though he paused briefly to survey the scene. Unsure of whether to tend to the wounded or to demonstrate his fortitude by entering the theater, he took Eugénie's lead and went inside to the imperial box, where they were greeted by a raucous ovation and shouts of *"Vive l'Impératrice"* and *"Vive l'Empereur!"*

The mastermind of the plot was an Italian named Felice Orsini. Arrested at the scene, Orsini was tried in February. He made an impassioned speech calling for Italian independence and denouncing Napoleon for betraying the cause of Italy, and was sentenced to death. His plot had repercussions. Napoleon had been ambivalent about using the police to suppress dissent, but the Orsini plot caused him to reconsider. It also propelled him to support Italian independence. He had spent much of his life fighting for lost causes, in Italy and in France, and he respected Orsini's passion. Within six months of the assassination attempt, Napoleon met secretly with Count Camillo Cavour of Piedmont and conspired against Austria, which was then in control of northern Italy. The brief war between France and Austria in 1859 was won by France and secured Italian independence under the rule of Piedmont and its king, Victor Emmanuel II. In return, France acquired Nice and Savoy.

But, more pertinent to Suez, the Orsini affair deepened tensions between Britain and France. Orsini had spent years in exile in London; the attempt on Napoleon's life was planned in England, and the bombs had been manufactured in Birmingham. One of the plotters, in fact, remained in London. The French press labeled England "a nest of vipers" and "a laboratory of crime," and the French government demanded the extradition of the plotter. Palmerston balked at this potential infringement on British sovereignty. French public opinion turned against the English, and there was talk of war. Palmerston's

enemies in Parliament, appalled at his cavalier attitude toward potential hostilities with France and delighted to use the crisis to undermine
his ministry, took advantage of the war scare and brought down his
government. Though the crisis blew over, Palmerston was out of office
for a year.[5]

Suez thereby received an unexpected boost from an unintended
quarter. With the emperor less willing to placate English concern
about French plans for a Suez Canal, and with Palmerston temporarily
removed, Lesseps suddenly found the field cleared of two obstacles.

That spring of 1858, he stepped up his already hectic pace. He was
in Constantinople twice, and he moved from one European city to
another in order to prepare public opinion for the funding of the company. Writing from the Ottoman capital, he told Negrelli that the time
was near. "In France," he said, "the opposition to England will be our
principal draw." And yet the refusal of the sultan or his ministers to
make a declaration in favor of the canal remained a serious impedi-
ment. Ali Pasha and Fuad were no closer to granting approval that
spring than they had been in the previous three years, and they were
masters of equivocation. Still, even though the Porte remained unprepared to give its official yes, it also remained unprepared to announce
an official no. In true cup-half-full fashion, Lesseps decided that, as
long as the Ottomans did not actively intervene, the project could
move forward. He continued to argue that the concession granted by
Said was sufficient, and that the approval of the sultan, while desirable, was not necessary.[6]

In May, Suez received yet another boost when the French Academy
of Sciences submitted its official report. Perhaps the most prestigious
of the many societies of scientists and engineers, the academy ratified
the plans submitted by Lesseps, Linant, and Mougel, and it saluted
Lesseps for his willingness to entertain all objections. The members of
the academy, who had once been skeptical that an amateur like
Lesseps could lead a complicated project like Suez, acknowledged that
he had proved himself to be a man of learning. Rather than resting on
one survey, or even two, he had authorized repeated fact-finding mis-
sions to the isthmus and presented his notes and those of the project
engineers to the scrutiny of all who might question the plans.

The academy subjected these documents to careful study, and its
members evaluated the testimony of engineers and men of learning
from more than seven different countries. It was not disposed to lend

its sanction promiscuously. It was composed of men who took their expertise seriously. In portraits, they looked somber, and exuded an aura of intellectual and cultural superiority, signaling that they considered themselves the guardians of standards. In a culture that genuflected to the wisdom of secular learning, their opinions mattered. And so it was a great victory for the canal when the academy concluded that none of the potential problems should stand in the way of commencing work. The report acknowledged the difficulty of constructing a port at Pelusium, and it cautioned that questions remained about the duration and intensity of the tides on the Mediterranean coast and in the Red Sea, as well as about the potential for silting on the one hand and innundation on the other. But it found that Lesseps had taken due consideration of these problems and had satisfied the academy that he would be able to solve them.

The report also noted that consuls and chambers of commerce throughout Europe had stated an interest in seeing the passage built, and it rebutted criticisms that the canal would be exorbitantly expensive. In the past two decades, Europeans had spent twelve billion francs constructing railroads. At two hundred million francs, therefore, the canal was a bargain, and would be even if the project went over budget. And as a body charged not just with upholding scientific standards but with the implied mandate to protect the interests of society and civilization, the academy saluted the moral implications of the canal: the shorter route to Suez, the report concluded, would "bring the nations of the world together for their common good."[7]

The academy report was the latest in a series of high-profile endorsements, and the English opponents of the canal recognized that Lesseps was gaining momentum. The best they could do was to prevent the government from completely caving. In June 1858, the Suez question was debated in the House of Commons in response to a resolution calling on the government not to use its influence to "induce the Sultan to withhold his consent to the project of making a Canal across the Isthmus of Suez." A heated argument ensued, with opponents declaring that the honor of England was at stake and that the seemingly innocuous resolution could lead to the rise of a new French Empire at England's expense. Robert Stephenson rehashed his old objections, calling the scheme technically impossible and politically unwise. Gladstone responded with an attack on the government for its stubborn intransigence against an undertaking that would benefit the commerce of England, and which all sane businessmen favored. He used the canal as an excuse for an eloquent assault on Palmerston, who, though out of office, still led his party and possessed immense influence. Gladstone's fervor notwithstanding, the resolution was soundly defeated. It had become an issue of English pride to oppose the canal, and though many favored its construction, few wanted to give the impression of bowing to French or international pressure.

Lesseps reacted to this debate with a combination of bemusement and indignation. He had given up on converting the British government, but he was bothered by the continued denigration of his integrity and that of the project. Knowing that many influential men of commerce in England wanted the canal built, he was disappointed that their sentiments did not translate into political support. He felt, with some justification, that the English opponents were misrepresenting the technical issues and the political ramifications. He also was losing patience with what he took to be the Hamlet-like indecision of the Ottomans. It had become clear to him that the only way to get things done was for decisive individuals to take action. "If every great improvement had to be suspended until it was sanctioned by some official authority," he wrote to a friend in 1858, "the world would be stopped in its tracks, or it would move backward."[8]

The years of carefully building international support for the canal were essential to its eventual construction, but at this juncture, Lesseps was thoroughly frustrated by the vacillation of the various parties. If the matter were left to the whims of Said, Palmerston, Napoleon, the sultan, and the ministers in Constantinople, Paris, London, Cairo, and Alexandria, the canal would remain a splendid idea whose time had not yet come. Though he was a former diplomat, Lesseps had remarkably little patience with the slow cadence of diplomacy. By the middle of 1858, he had assembled everything he needed to begin the work—everything, that is, except two hundred million francs. He had the tacit approval of Napoleon III, and the active support of public opinion in France and even in England. It was time, therefore, to raise money and get to work.

# CHAPTER ELEVEN

# A UNIVERSAL COMPANY
# FOR A MARITIME CANAL

O N OCTOBER 15, 1858, Ferdinand de Lesseps released two letters. One was sent to members of the press throughout Europe;
the other was addressed to accredited agents of the Suez Canal
Company. In both, Lesseps announced that, on November 5, four hundred thousand shares in the company would be offered to the public at
a cost of five hundred francs per share.

Both circulars sketched the outlines of the project. Following the
statutes drawn up by Lesseps and Said in 1856, the company was to be
governed by the shareholders. Its purpose was to construct a canal
suitable for ship navigation between the Red Sea and the Mediterranean, and an irrigation canal from the Nile to the isthmus. The company was granted rights to all land cultivated by this irrigation canal.
The cost of construction was estimated at 160 million francs—though,
once interest charges payable to shareholders were factored in, the
total cost would equal the two hundred million to be raised from the
subscription. Lesseps promised that the anticipated revenues more
than justified the expense. Once the canal was complete, at least three
million tons of goods would pass through the canal at a charge of ten
francs per ton, for an annual revenue of thirty million francs. That figure, Lesseps claimed, was conservative and represented only a modest
percentage of the shipping that made the voyage around the Cape of
Good Hope. The work was expected to take six years. First the freshwater irrigation canal would be dug. Then, once that was complete and
lands were brought under cultivation, revenue would start flowing to

the company. After two more years, a rudimentary canal would exist between the two seas, and the rest of the work could then accelerate.

As even Lesseps suspected at the time, the estimated time-frame was wildly optimistic, but how else could he have generated the enthusiasm? No entrepreneur launches a bold and speculative venture by announcing to the world that the work will take longer than anyone imagines, that the cost will exceed original estimates, and that shareholders will see their initial stake diluted and their prospects for profit dwindle. In all speculative endeavors, there is a dance of seduction, a wink and a nod that, though it would be wonderful if the best-case scenario proved correct, there is precious little chance that it will. Most investors, much of the time, understand the risks, or they allow themselves to be lulled by the unlikely possibility that the best-laid plans will, in fact, result in near-perfect execution.

Just in case the financial prospects were not appealing enough, Lesseps dangled one other attraction. Purchasing shares, he said, would be a contribution to the onward march of progress. It was not unusual for entrepreneurs in the mid-nineteenth century to cloak what might otherwise have been prosaic engineering projects in larger terms. That was part of the reason such projects were funded; making money was another. For five hundred francs, anyone anywhere in Europe could contribute to the betterment of mankind, and create a better future for themselves and their families. Those who wished to join this endeavor were instructed to go to their local bank, place a fifty-franc deposit against each share, and then inform Lesseps personally of the name of the banker and the name of the shareholder. The next fifty francs would be due once the initial sale was approved at the company's headquarters in Paris. The schedule for the remainder of the balance would be published soon after.[1]

Arrangements had already been made with the larger banking houses, and leading financiers were given blocks of shares in return for their assistance. Money would have to be collected from hundreds of small local banks, and then funneled to the company. The process was complicated, but though Lesseps needed to work with bankers, they were merely his clients. As he had promised Baron Rothschild, Lesseps had created the company on his own.

It was not as if these two letters were sprung on the public without prior planning. Already, every significant city in Europe and Britain had an agent of the company, with an office and a budget. These agents had spent many months meeting with business leaders and the local aristocracy, and wooing the members of the press. By the fall of 1858, the project was known and understood throughout the continent, and Lesseps had every reason to be confident that the shares would be bought, and bought quickly.

But he was not prepared to sit and wait. For months, he had been attending banquets and receptions in various cities, and he continued his pace throughout the fall. For years, the canal had been billed as an international effort and not simply a French one. Though he had lobbied intensively in England and Scotland, it was uncertain how many shares the English and the Scots would purchase. It was even more unclear how many people outside of Britain and France would participate. Yet, in order to support the assertion that the venture was dedicated to the good of all nations, the company had to consist of international shareholders. In fact, to assure Said (who was deeply worried that he would look like a French stooge if the only shareholders were French), Lesseps set aside eighty-five thousand shares for sale in England, Austria, Russia, and the United States.

He then went on a road show. In August, he visited Odessa. Though it was a major Russian city, it was not exactly a cosmopolitan European center. Honored by the visit, the city dignitaries lavished Lesseps with praise and treated him to a banquet on a yacht called the *Vladimir.* Attended by luminaries such as Count Stroganoff, the governor-general of Bessarabia, and the director of the Imperial Bank of St. Petersburg, the feast concluded with a fulsome toast lauding the canal as an achievement that marked the apex of human civilization and which would, in its way, reflect glory on both Lesseps and Tsar Alexander II of Russia. Lesseps thanked his hosts and compared his project to Russia's Company of Steam Navigation for the Caspian Sea, and to the waterways that Russians had constructed to enable more commerce on the Volga and the Don. Further toasts followed; vodka was consumed; and the rhetoric soared to heights only attainable in the afterglow of too much food and just enough alcohol.

Similar scenes were replayed in Trieste, Vienna, Barcelona, and Turin, as well as in French cities such as Bordeaux and Marseille. Lesseps addressed the Chamber of Commerce in Barcelona, a city he had once called home. Businessmen, nobles, men of science and culture, and more than a thousand spectators crowded the hall to hear Lesseps explain, in Spanish, why the time had come to build the Suez

Canal. Reporting on the event, the main journal in Barcelona explained why the canal was imperative: "Turkey is troubled; India is trying to rise up against the domination of a sole power; Christian civilization is knocking on the door of China and Japan; it is urgent that there be a shorter route between the Occident and the Orient to speed up interactions between their peoples. History tells us that God is always on the side of those who have the means to satisfy great and legitimate needs." In Turin, a newspaper reported that banks were preparing for heavy demand on November 5, and that the people of the region eagerly awaited the day. The promise of an "army of workers and numerous machines removing the soil of the isthmus and then, by virtue of the irrigation canal, making fertile the ancient land of Goshen, which was heralded in the blessed Scriptures. . . . Those who are timid and uncertain about the future should remember that European science will transform Suez into a new Bosporus, a new route to enrich the world to parallel that ancient route that had enriched Rome and Constantinople."[2]

By late fall, Lesseps believed that he had effectively made his case, that shares would be bought throughout Europe, and that the enthusiasm of those many banquets and late-night toasts would be translated into bank deposits and agreements to buy shares.

When the subscription closed on November 30, however, the results were less than stellar. Slightly more than 23,000 people had purchased shares, but over 21,000 of them were French citizens. All in all, French shareholders held more than 200,000, or just over half, of all the shares offered. The next-largest block was bought within the Ottoman Empire, including Egypt; the personal stake of Said alone amounted to more than 60,000 shares. Another 4,000 were purchased in Spain; 2,600 in Holland; and 1,300 in Piedmont. And as for those 85,000 shares reserved for Great Britain, Russia, Austria, and the United States, none were bought. The verdict was clear. Ferdinand de Lesseps had sold his vision to 21,000 Frenchmen and one prominent Egyptian. The rest of the world, however favorably disposed, preferred to stand on the sidelines, cheering perhaps, but not risking money.[3]

The popular enthusiasm in France was good for the canal in all but one respect: it made the lukewarm reaction elsewhere seem more negative than it was. Given the cost of the shares, the response of the French was more extraordinary than the indifference of the rest of Europe. Five hundred francs was not an inconsiderable sum in 1858. It was slightly more than the average annual income in France. The shares sold for twenty pounds in Britain, and that was more than a member of the working class could afford. In addition, though it was possible for an individual to buy only one share, the company's agents insinuated that only those who took at least ten would find their requests honored. Here, the response was tepid. In the end, many people owned fewer than ten shares. They were willing to risk five hundred or a thousand francs, but five thousand was beyond the limit most investors wanted to go. Farther east in Europe, the pool of potential buyers was smaller. In Austria, Russia, and Spain, only a tiny portion of the population had the resources to become shareholders, and in the Ottoman Empire, only the thin layer of the ruling classes had the ability and the wherewithal to invest. In the middle of the nineteenth century, the joint stock company was still an anomaly outside of Britain and France; the very idea that thousands of people could have a stake in an engineering project would have been alien only a few decades before. The railroads of the era were being built on a similar principle, but, historically, major public works had been paid for by the state, either nationally by the king or locally by nobles and municipal grandees.

The novelty of a "universal company" only partly explains the indifferent response outside of France. In spite of the efforts of its promoters, the canal was still viewed as a French project overseen by French businessmen to serve the national interests of France. Though Russian aristocrats and the minor nobility of Vienna looked forward to the canal's completion, they refrained from investing. They saw no reason to expose themselves to potential loss, even if they encouraged Lesseps to pursue his dream. If he succeeded, they would all benefit. If he failed, at least they would not go down with him.

Lesseps had difficulty accepting the reality of what had transpired. In an act of questionable accounting, the company treated the shares that had been reserved in advance as if they had actually been purchased. In a sharp exchange with Baron Bruck, the Austrian finance minister who had been closely involved with the preliminary organization of the company in Vienna, Lesseps demanded that five million francs be paid to the company. His rationale was that the Austrian agents had promised that fifty thousand shares would be bought, and Lesseps considered that a binding commitment. Bruck and the Austrian representatives took the more reasonable stance that a promise to sell fifty thousand shares could not be construed as a legal contract to pay for them. In any event, where was the money supposed to come from? The shares had not been sold. If Lesseps demanded an indemnity, who would indemnify him? Here, as in many of the disputes that surrounded the canal's creation, Lesseps attempted to turn kind words into financial commitments. That did not work with the Austrian Empire.[4]

It did, however, yield results with the ruler of Egypt. Said's reasons for supporting the canal in November 1858 were no different from what they had been in November 1854. The difference was in scale and in scope. The first commitment had entailed no more than thirty thousand francs a month to fund propaganda and preliminary studies—not a pittance, but a far cry from the eighty-eight million francs he would now contribute. That was more than the total annual revenue collected by Said's government, and though the amount would come due in stages over the course of several years, it was a huge financial obligation. Said initially agreed to purchase sixty-four thousand shares, but he also promised Lesseps that the Egyptian government would acquire whatever shares were outstanding once the subscription closed at the end of November. Trusting Lesseps's assurances that the initial flotation of four hundred thousand shares would succeed, Said imagined that at worst he would have to cover the balance of a few thousand outstanding shares. When none of the eighty-five thousand shares set aside for other countries were bought and tens of thousands of other shares remained unsubscribed, Lesseps turned to Said and requested that the ruler honor his promise.

Said could have refused, but he believed that his personal prestige as well as the honor of Egypt were at stake. Even though Lesseps encouraged this perception, no matter how good a Svengali he was, he cannot be made to shoulder the entire blame for Said's decision. Subsequent generations of Egyptians and Europeans characterized Said as a rube who was manipulated to underwrite the canal, but the reality is more complicated.

The years had worn on the relationship between Said and Lesseps, and the viceroy no longer looked at the Frenchman as an old childhood friend. He had come to realize that Lesseps could and would discard anyone who stood in his way. That must have intimidated Said, who, much like Napoleon III, disliked confrontation. At the same time, Said could have said no. That would have infuriated Lesseps, and the company might have been able to coerce him to pay an indemnity. Even worse, his refusal might have doomed the canal. The statutes demanded that the company be fully capitalized. Unless all the shares were subscribed, it could not be duly constituted as a legal entity and would be unable to collect the money due on the shares. Without Said, the entire project could collapse. He did not want that to happen. He had decided that the canal served the interests of Egypt and his family, and though he may not have appreciated being presented with a bill for the outstanding shares, he accepted the burden. Better to invest more in an actual canal that he believed would benefit Egypt than to be forced to pay millions to compensate a group of French entrepreneurs.

As it turned out, the final tally for Said was 177,000 shares, which included the original sixty-four thousand to which he had committed himself, plus all the unsold shares from the end of November. Some accounts say that Lesseps demanded payment and addressed Said like a father berating a recalcitrant child. Some say that Said angrily resisted, only to be subjected to veiled threats from Lesseps that, whereas the future of the canal was certain, Said's was not. Still others suggest that Lesseps arranged a secret meeting with Napoleon III and then demanded payment from Said in the name of the emperor. But all of the stories smack of later propaganda, designed to bolster arguments that the canal was an albatross foisted on Egypt by an avaricious Lesseps.

In truth, Said may have resented the burden, but he also shared the dream. And, much like Lesseps, he wanted his name to be immortalized. "People are mistaken," Said told the British consul, "if they attribute the piercing of the isthmus to Monsieur de Lesseps alone, for I am a promoter of it. Monsieur de Lesseps has merely carried out my instructions. You will ask me perhaps what my motive has been, and I will tell you that it has been to bring honor to my name and serve at the same time the interests of the Ottoman Empire." Inheritor of an ancient legacy of Egyptian autocracy, Said never strayed from the vision spun by Lesseps that day in the desert years before. Even as his friendship with Ferdinand deteriorated, they were bound by a common goal.[5]

On December 15, having secured Said's support, Lesseps announced the official formation of the Suez Canal Company. That in turn led to the first meeting of the company's governing body, and the first meeting of the administrative arm charged with overseeing the actual physical work. Billed as a universal company, it was primarily a French-Egyptian company with a smattering of others, but the Administrative Council included Parisian bankers, Belgian public-works officials, a Russian steamship entrepreneur, the inspector general of the Orléans railroad, the head of the Venetian Chamber of Commerce, and Paul Forbes, a banker from Boston. The company was structured to give Lesseps ultimate say on all matters of importance, subject only to annual approval by a meeting of the shareholders.[6]

At no point did Lesseps face serious opposition from the thousands of people who had invested their money in his spectacular adventure. Though the statutes gave him control, he rarely needed to invoke his authority against the will of the shareholders. They had invested in the canal because they had been swept up by the vision spun by its founder. They revered Lesseps as a man who would not only make them rich, but was offering them a precious opportunity to participate in the march of civilization.

For the most part, these investors were not prominent members of society. They were middle-class, conservative bourgeois who had earned their money through cautious work and were respected members of their communities. They were scorned by the British press, who dismissed them as ignorant villagers duped by a charlatan. According to *The Times,* naïve French investors were burying "their savings in the mud at the mouth of the Nile. . . . Were not de Lesseps' scheme the merest moonshine the share in his company would be bought up in twenty-four hours by English shareholders." In the words of another London paper, the investors were mostly "hotel waiters, who have been deceived by the papers they had read, and petty grocery employees who have been beguiled by puffs. The priesthood has been victimized. . . . The whole thing is a flagrant robbery gotten up to despoil the simple people." Palmerston, surveying the list of those who had bought shares, was unable to resist yet another gibe and scoffed, "Little men have been induced to buy small shares."[7]

Though the condescension revealed more about British animosity than about French society, there was some truth in these characterizations. Very few individuals had bought large blocks of shares. In fact, only 128 had purchased more than a hundred shares, and only sixty had taken more than 150. The average holding was nine shares. The investors formed a remarkable cross-section of French society. According to the list published by the company, there were 91 mechanics, 249 engineers, 267 judges, 369 bankers, 433 doctors, 434 teachers, 480 priests, 819 lawyers or notaries, 928 artisans, 973 soldiers, 1,309 civil servants, 2,195 clerks, 4,763 merchants and industrialists, 5,782 land proprietors, and another 4,000 classified as unknown or miscellaneous. The thriving industrial cities of Bordeaux and Lyon took the largest block outside of Paris, followed closely by Marseille, a port city that stood to benefit greatly from the canal.

The shareholders were professionals who formed the core of mid-nineteenth-century France. They were a product of the industrializing world of the Second Empire, earnest, hardworking men who drew the disdain of the same intelligentsia who reviled Napoleon III. They valued patriotism, thrift, and hard work. They wished to make money and acquire nice things. They deferred to authority, respected hierarchy, and rallied to the flag. They believed that society should be ordered from top to bottom, with children deferring to parents, wives to husbands, clerks to employers, all to the emperor, and the emperor to God. They supported Napoleon III, and craved the stability he offered. They were at once very French and very much like their counterparts throughout the world. The intelligentsia thought them terribly boring and possibly inimical to national greatness. But without them, society tends to break apart into bitter factions of ideological absolutism.

It was their imagination that Lesseps had captured, yet they were a class routinely dismissed for their lack of imagination. Writers such as Hugo, Zola, and Balzac took meticulous pains to skewer them, and literary and artistic culture in Paris organized itself in part around the principle of *"épater la bourgeoisie"*—namely, stomping on the sensibilities of these bankers and doctors and lawyers who formed the acquisitive backbone of society. That tone became even more prevalent later in the century, but it was already apparent during the Second Empire. But for all of their scorn, the literati never could explain how it was that these supposed dullards were willing to risk their hard-earned cash on the filigree promises of Ferdinand de Lesseps.

The bourgeoisie yearned for stability, but they lived in a culture that swirled with romantic images. Literate and engaged in the world of affairs, they could not avoid the whirlwind of new ideas. They may have rejected Enfantin and other outlandish prophets, but they were influenced nonetheless. As conservative as they were in their sense of how society should be ordered, they lived in a sea of conflicting doctrines. Progress battled with a Catholicism resurgent after decades of attack. Theologians argued with secular rationalists such as Ernest Renan, who, having nearly become a priest, instead devoted himself to pillorying traditional religion. Science versus religion was only one of many battles. Demands for the education of women jockeyed with jeremiads that women were becoming too liberated. And the march of industry was greeted in many rural areas as a threat to a way of life that had survived wars and revolutions yet seemed defenseless in the face of the railroad and the telegraph.

But society was also suffused with romantic images in a much more literal sense. The passion for things Egyptian unleashed by Napoleon's occupation of Egypt never waned. After Champollion and the cracking of the hieroglyphic code, it deepened, and the contest with England over the fate of Muhammad Ali kept France focused on Egypt. Fascination with the Orient was not particular to France; nineteenth-century Germans, Americans, and English had a similar mania. But French "Egyptomania" was more intense. From the time of Bonaparte at the beginning of the nineteenth century, French painters took trips to North Africa, to Turkey, to the Holy Land, and to Egypt. They absorbed what they saw, and transformed those images into paintings. Writers traveled as well, and turned what they saw into poems and novels, which then in turn became source material for painters. By midcentury, a new medium had been added, photography. Combined, these photographs, paintings, and books fueled the passion of the French for the East. Without it, the selling of the Suez Canal would have been far more difficult.

The old conflict with the Ottomans, and the rivalry between Islam and Christianity that flared most notably during the Crusades, were part of the romance. In the early eighteenth century, a Frenchman named Antoine Galland published a translation of *The Arabian Nights*, and European audiences read with glee the titillating tales of Scheherazade and Caliph Harun ar-Rashid cavorting with geniis and bandits in medieval Baghdad. The stories of Sinbad were not much different from Homer's tale of Odysseus, but educated Europeans were thrilled by the exotic, and erotic, harem. The image of eunuchs guarding the entrance to a world of sexually available women lodged in the collective imagination of Western Europe, and there it remained for the next two hundred years.[8]

Painters had been using the Bible as a primary subject for centuries. In that sense, attention had always been focused in the Orient. But in the nineteenth century, as religion lost its cultural centrality, painters looked for new subjects in old places. The invasion of Egypt and the subsequent French occupation of Algeria opened these areas to travelers. Even those painters who did not themselves go for extended stays in Algeria, Morocco, Istanbul, or Cairo were influenced by those who did, and when photographs of these areas began to trickle forth in the 1840s, those too were incorporated into paintings.

Some of the most illustrious artists of the age seized on Oriental themes and combined Biblical stories with myths and legends. Eugène Delacroix composed one of his most famous images of the East before he actually went there, and used a poem by Lord Byron as inspiration. The exhibition of his *Death of Sardanapalus* in the Paris Salon of 1827 was a happening, widely discussed and debated. The painting flowed with naked women surrounding a dark and brooding potentate in a swirl of sex and violence. Later, after a stay in North Africa, Delacroix painted the equally famous *Women of Algiers,* which depicted three women in their quarters, unveiled, relaxing, dressed in fine garments. While critics argued over the painting's merits, it unleashed a cascade of discussion and imitation.

It was said of Delacroix (by Baudelaire, no less) that he "was passionately in love with passion." But Delacroix himself was influenced by the no less passionate poetry of Victor Hugo, who published *Les Orientales* in 1829. Hugo was the son of one of Napoleon's generals, and he had studied at the Polytechnic School. Possessed of more than his share of self-assurance and without any direct knowledge of the Near East, Hugo spun a web of magical poems. He emulated Scheherazade. He described the winds rolling off the Sahara and the mysteries of ancient Egypt then being revealed by French archeologists, and he evoked a decadent East that held both promise and peril.

One of Delacroix's rivals and critics was Jean-Auguste-Dominique Ingres, whose canvases were more controlled and less flamboyant. Ingres also peered into the harem and tried to expose the sensual pleasures hidden inside. Whereas Delacroix was all color and action, Ingres presented meticulous set-pieces, frozen in time. But he was no less entranced by the Orient. He painted "odalisques," an archaic word for concubines or harem slaves. His odalisques, statuesque and pale, were graced with translucent skin, and their eyes gazed off, as if they were waiting, not altogether happily, for their lord to arrive for his evening pleasure. In *Odalisque with a Slave,* which Ingres finished in

1839, the woman lies naked, stretched out on a divan as her female servant plays an instrument and a turbaned Nubian stands guard. The setting is ornate and mournful; the mood is expectant.

Ingres was not simply one of the most successful painters of his day.
He was also an arbiter of taste in a world where success as an artist
depended on the validation of the Salon, and he was a teacher who
schooled others in his craft. Because he enjoyed a place at the pinnacle
of the European artistic establishment, his images were not just widely
viewed and debated but influenced other, younger painters to take up
similar subjects. Soon, images of the Orient became almost commonplace. At any given time, artists tend to gravitate toward similar subject
matter, and by the middle of the nineteenth century, dozens of artists
and photographers looked to the Near East and North Africa for inspiration. Robed Bedouins, camel caravans, Cairene streets, ruined temples, scenes of tent life replete with sumptuous meals and dark
interiors, and the ubiquitous harem guarded by muscled, dark-skinned
eunuchs cropped up on canvas after canvas.

The Swiss painter Charles Gleyre, the Scottish draftsman David
Roberts, and the French artist Horace Vernet created hundreds of pictures of the Orient. The prolific Roberts, whose drawings and prints
were hugely popular in England and on the continent, captured both
daily life in Egypt and the ancient temples that lined the Nile. The
elaborate paintings of Gleyre and Vernet steered away from mythic
narratives and instead depicted everyday life in the Near East. The
style was dramatic, but the subject matter was not. Vernet painted a
canvas showing an Arab storyteller entertaining a small group relaxing
in a desert oasis. Gleyre composed portraits of Egyptian peasants going
about their day. Prosaic topics, but for a banker in Lyon, or a clergyman
in Bordeaux, it was all deliciously foreign.[9]

Enchanted by these images, two young men went to Egypt in 1849.
One was a rather small, brooding twenty-six-year-old writer and pho-
tographer named Maxime Du Camp. The other was a writer and an
aesthete, gregarious, twenty-eight, tall, and blond, who had read Hugo
and Byron and *The Arabian Nights;* had penned a novel about St.
Anthony's temptations in the Egyptian desert; and now longed to
stand at the foot of the Great Pyramid. "Oh, how willingly I would give
up all the women in the world," he wrote in one of his more sober
moments, "to possess the mummy of Cleopatra!" His name was Gustave Flaubert, and shortly after arriving in Egypt, he and Du Camp

decamped to Cairo via a boat on the Nile. Falling asleep while gazing at the brownish water, Flaubert noted simply, "Such rapture!" Observing the world around him, he wrote, "Here, the Bible is a picture of life today." It was in many ways better than he had imagined. Once in Cairo, and ensconced at the Hôtel du Nil, he complained that there were no longer any good brothels in the city and that all the dancing girls had somehow relocated to Upper Egypt. He managed to find women nonetheless, and he graphically described his many couplings in letters home.

Flaubert was most entranced, however, by a woman who remained unavailable to him—the proprietress of his hotel, Madame Bouvaret. And one day in the spring of 1850, while traveling in the desert in Upper Egypt in the region of Aswan, he had an epiphany for the name of the main character of a story he had been mulling, a story about a sexually liberated woman. Her name, he exclaimed, would be "Bovary! Emma Bovary." Out of that mélange of heat and sex and sand was born one of the most famous characters of nineteenth-century literature.[10]

In letters and subsequent writings, Flaubert popularized an exotic portrait of Egypt. The photographs taken by Du Camp on this trip were published and widely disseminated. But life in Egypt was not nearly as exotic for those living it as it was for European visitors observing it. In fact, even the harem was more mundane than what the artists who had never seen the inside of one imagined. As one European woman who spent time visiting a harem observed, there was not much sex and little naked lounging; there were, instead, many cups of coffee and lots of talk about fashion, making the harem not so different from equivalent social scenes in London or Paris.[11] But that is not how most Europeans saw Egypt and the rest of the Orient.

As artists and writers offered new windows into these faraway lands, the public became hungry for antiquities, both for objects themselves and for stories about them. From the curious schoolboy to the secretive Masons, many felt the magical allure of ancient Egypt. One of the heirs of Champollion was the young archeologist Auguste Mariette, who arrived in Egypt in 1850 at the age of twenty-eight. He had read all that he could about Egypt, and, having exhausted his book knowledge, he set out on a parcel-post boat from Marseille and headed for Alexandria. Overcome by the heat and the noise and the new smells that met him off the boat, he went on to Cairo and arranged a meeting with Linant Bey, who at the time was out of favor with Abbas Pasha

and living in an old mansion with his Ethiopian wife. Linant then introduced him to others, and though Abbas had banned European archeologists, Mariette began a clandestine excavation around the Pyramids.

Mariette was perhaps the only person in Egypt in these years who
rivaled Lesseps as a self-promoter. In fact, he used Lesseps to advance
his own career. Before leaving for Egypt, Mariette had been told by one
of Muhammad Ali's French civil servants that, though the prospects for
the canal were doubtful, it was impossible to rule out the possibility.
"The Orient is the land of images and miracles," he was told. "Often,
what seems the most improbable happens." Mariette took this dictum
to heart. In 1855, he convinced Lesseps to introduce him to Said, and
he then convinced Said to pay for his work and make him director of
antiquities. So began a long career in Egypt, one that would frequently
intersect with the canal. At several junctures, Mariette helped promote
the canal, and in return Lesseps helped ensure that Mariette's budget
would not be cut. The relationship served both men well: the canal was
constructed, and Mariette became one of the most influential archeologists of the day.[12]

Without the legacy of these writers, artists, photographers, and archeologists, Ferdinand de Lesseps might not have been able to convince
those thousands of artisans, bankers, clerks, doctors, and lawyers to
invest in the canal. One investor wrote Lesseps claiming that, though he
had never put money in a speculative venture, he was proud to buy a few
shares of the Canal Company as testament to his loyalty to France. But
in all likelihood, he and many others were also motivated by reasons that
were hard to explain and uncomfortable to admit.

Without question, shareholders were drawn to the idea that money
invested in the canal was money used for the good of France and for
the good of civilization. The prevalence of "Oriental" themes in France
at midcentury made Egypt and Suez seem closer and more familiar,
and that in turn probably helped the cause of the Canal Company.
Egypt was far away, but it was also everywhere, in books, exhibitions,
poems, and journals.

Underneath the surface of everyday life, there was another reality,
one that Lesseps himself may have been oblivious to but which pulled
investors toward Egypt as much as the Canal Company's artful propaganda. Society pulsed with images of the Orient, and these acted on
people's hidden desires. In ways that can never be documented, the
mystique of Egypt exerted a powerful pull. It doesn't matter how much

those images corresponded to reality. It matters how much they predisposed the public to be seduced. Those bankers and lawyers were portrayed as lacking in imagination, but, like the upright Victorian gentlemen with a healthy, yet secret, appetite for erotica, they had their private longings, and these had outlets. Having been enchanted by lascivious harem scenes, tales of caliphs and maidens, pictures of a life in the desert unfettered by the rules of modern society, and by the mysteries of the ancient artifacts slowly being uncovered in the desert, these doctors and clergymen and clerks saw their shares in the Suez Canal Company as much more than a patriotic investment. They were shares of passion.

CHAPTER TWELVE

# The Work Ahead

ENTRUSTED WITH THE money of thousands of Frenchmen, hundreds of Turks, Spaniards, Dutch, Austrians, and Italians, and one anxious ruler, Lesseps turned to the next task at hand: parting the desert and uniting the two seas.

As of January 1859, a company existed. Its primary offices were in Paris, at 12 Place Vendôme, and in Alexandria. It had money from the newly purchased shares. Though the British government and much of the English press were hostile, the attitude in France could hardly have been more supportive. "This enterprise," claimed an editorial in *Le Siècle,* "will bring eternal honor to M. Ferdinand de Lesseps." Lesseps, in a formal letter to Said Pasha asking for authorization to commence work, wrote as if the outstanding issues had all been resolved. True, the sultan had not given his consent, but Lesseps did not waver from his position that such approval was not legally necessary. After all, he reminded Said, as part of the landmark Hatt-i Humayun decree of 1856, the sultan himself called for the improvement of land and sea communications throughout the empire and specifically mentioned canals. Therefore, even though the sultan had not yet formally endorsed the Suez Canal, he had, according to Lesseps, endorsed it in principle.

As always, Lesseps was deferential in his correspondence with Said, but though he wrote as though he was making a request, he was actually making a demand. According to Lesseps, Said's prior decrees were tantamount to an inviolable obligation, and Said's desire to get permission from the sultan had no bearing on the rights of the company. If Said felt that he needed the sultan's blessing, then it was his responsibility to obtain it. In the meantime, the company's board of directors had met; its statutes were formalized; and Lesseps intended to begin.

Yet, even without the political obstacles, there was a gap between what Lesseps wanted to do and what could actually be done. The canal's partisans had spent four years lobbying for support and establishing the company. Surveys had been conducted; good will had been generated. But Port Said, Suez, Lake Timsah, and the isthmus looked only marginally different in 1859 than they had in 1854. The region was still desolate, uninhabited, and arid. A few huts on the strip of sand optimistically named Port Said marked the extent of the work to date. As far as the canal project had come, it remained a paper plan.

In fact, it was worse than that. There was no corporate structure to oversee the complicated task ahead. The existing surveys of Linant, Mougel, and others were useful as primers but inadequate as actual blueprints. Lesseps could direct the project, but he was neither a contractor nor an engineer. In short, the Suez Canal Company had been formed to execute an extensive and challenging endeavor, yet the only thing it was capable of doing at this point was issuing shares, and it didn't even do that very efficiently. In the first months of 1859, Lesseps was barraged with anxious letters from shareholders who complained that they had sent their money but had received no reply, and he was reduced to writing personally to assure them that, though there had been some delay, the certificates were being printed.[1]

Meanwhile, at the end of 1858, the company's Works Committee convened for the first time. Lesseps had managed to assemble an impressive collection of engineers from across Europe, and though he nominally had the final say, he knew better than to insert himself too aggressively into the deliberations. Instead, Mougel Bey led the meetings. The simmering tensions between Linant and Mougel had finally crested, and Mougel emerged victorious. Linant, who had been in Egypt longer than Mougel, claimed that he had worked on the canal idea first and that he deserved precedence. Mougel, though junior, had more than twenty years of experience in Egypt. Each had a valid claim to pre-eminence, but only one could be appointed chief engineer and director of works. Forced to chose, Lesseps went with Mougel. Angry and bitter, Linant received a generous payoff from the company.

The Works Committee analyzed the job ahead and calculated how much the actual digging would cost. The initial focus was on the freshwater canal and on the area around Port Said and Lake Manzala, where the first stages of work would be concentrated. The most urgent requirements were to construct huts around Port Said to house two to three thousand workers and to arrange for the transportation and production of drinking water so that these workers could survive. There was no potable water in the vicinity, and the fishermen who trawled Lake Manzala relied on primitive distillation. The committee decided on a two-pronged approach to the problem: one was to build condensation plants, and the other was to transport the water by barges and by caravans from the Nile at Damietta. The other immediate need was to obtain stone for the Port Said jetties. One possible source was a quarry near Alexandria; there were also quarries a hundred miles south, along the isthmus, at Gebel Attaka. Both were to be explored.

Finally, a schedule was drawn up. The work was expected to take six years. Once the freshwater canal was completed, there would be a reliable and consistent source of water, and the number of workers could be increased to thirty thousand. Then a small access canal, eight meters wide, would be dug between Port Said and Lake Timsah. The access canal, the *rigole,* would allow the engineers to test the flow of the Mediterranean inland. If all went well, work would continue in the north on the larger, maritime canal, while to the south another access canal would be dug, between Lake Timsah and the Gulf of Suez, which would require blasting the Chalufa ridge, south of the Bitter Lakes. The jetty would be completed at Port Said; the canal would be widened and deepened around Lake Manzala, and then widened and deepened south of Lake Timsah. And then it would be done, at a total budget of two hundred million francs.[2]

The final report was an exercise in fantasy and wishful thinking. The committee used existing canals and jetties in Europe as benchmarks, but even if there was some precedent for the Port Said jetty, there was none for a canal of this scope and through this terrain. The financial projections were guesswork predicated on the assumption that there would be no significant delays and that the engineering challenges were understood. In truth, they were not.

Whereas the opponents of the project routinely overstated the obstacles, Lesseps and his partisans went to the opposite extreme. Mougel Bey, with his decades of experience in Egypt, should have known better, but perhaps sensing the desires of Lesseps, he went along with the sunny assessment. Conveniently overlooked in these initial meetings were the challenges posed by the three plateaus along the route. Though neither the Chalufa ridge in the south nor the Serapeum between Lake Timsah and the Bitter Lakes rose more than fifty feet above sea level, there had been no serious consideration about how to excavate them. The El-Guisr plateau in the north, located between the marshy bed of Lake Balah and Lake Timsah, would prove even more intractable, not because the region was rocky, but because it consisted mainly of sandy dunes. As anyone who has spent time at the beach knows, sand tends to shift in amoebic fashion, flowing back into the spot that has just been exposed. No one had addressed this dilemma, and the experience of European engineers with the moister, denser soils of Europe would be of little use.

But ignorance proved to be a blessing. Rather than focusing on the problems, Lesseps galvanized the staff of the company to find solutions. At each stage of the construction, as new difficulties revealed themselves, the engineers had no choice but to think on their feet and come up with solutions. Too much money and effort had been invested to allow the work to halt. Had the technical obstacles been fully appreciated at the outset, cost estimates would have ballooned, morale would have sunk, and the entire venture might have been fatally damaged.

In retrospect, what is most striking about the initial blueprints is how much the actual canal departed from them. Though the general route remained the same, the way the canal was constructed underwent a radical revision. For the first four years, the work relied on human labor, assisted by a few machines. In 1859, the company thought that the canal would be excavated in much the same way that public works in Egypt had always been completed, by the sweat and toil of hundreds of thousands of laborers. Said had agreed to furnish the company with Egyptian peasants, fellahin, who for centuries had been dragooned to work on irrigation projects or to erect the mausoleums and temples of the pharaohs. The modern Suez Canal may have been the brainchild of a nineteenth-century Frenchman, but it was to be built no differently from the canals dug by the Pharaoh Necho or by the Romans thousands of years before.

The preliminary plans did call for mechanical dredgers to deepen the channel through Lake Manzala, but for the most part the canal was treated as a straightforward task to be conducted in primitive fashion by unskilled workers with picks and baskets. By the standards of the day, the plans were unremarkable. Only in the past few decades had

Europeans and Americans started to use steam machines to do what human labor had done for all of recorded history, and the reliance on human labor to construct the canal struck no one as odd.

But halfway through, the plan was radically revised. Initially, the canal was an exercise in logistics—moving men to distinct points along the isthmus so that they could move earth. Yet, by 1863, these laborers had only managed to complete a fraction of the work, and when Said died suddenly and his successor halted the supply of fellahin, the future of the canal looked bleak. But this turn of fate proved to be a blessing. Had the company continued to rely on manual labor, the project might have stretched out for many more years and taxed the patience of its shareholders, even if the technical issues had been resolved with a few tons of dynamite and thousands of bodies toiling away. Instead, the prospect of a labor shortage forced a rethinking, and in 1864 Lesseps and his engineers turned to machines.

Mechanization transformed the project. The majority of the actual excavation and the construction of the jetty were done not in the first seven years of the company's existence, but in the final three years before the canal's completion in late 1869. Suez was supposed to link two worlds. It did, but in ways that Lesseps, the Saint-Simonians, and others had not foreseen. The canal did not just join Europe with the Orient; it also connected the preindustrial world to the mechanical era. For the first phase of its construction, it was a reflection of the past; during the second, it became a harbinger of the future.

Manual labor began the endeavor, and steam-powered engines finished it. In many ways, the two stages of the canal's creation mirrored the widening gulf between East and West. Until the industrial age, daily life in France was not too different from daily life in Egypt, India, or China. Language varied, as did religion. Specific history was distinct, as were geography and climate. But the needs of everyday life and the structures to supply them bore striking similarities across cultures and continents. Historians have argued, and will forever, over when the West began to diverge from the rest of the world, but there can be no question that industrialization was a physical manifestation of how different, and how dominant, Europe and North America were becoming. The mechanization of the canal's construction was a potent example of how the industrial age changed the nature of work and allowed Europeans and Americans to alter the shape of the earth more quickly and dramatically than any humans ever had.

But on that windswept day in April 1859 when Lesseps and his cohorts gathered at Port Said to plant the Egyptian flag and strike the symbolic first blow, they held pickaxes. Had they seen into the future, they might have stood next to a mechanical dredger, and rather than raising their axes to strike the sand, they would have gathered around Lesseps while he pushed a button and listened to the roar of the motor. As it was, work began with a simple blow struck by a tool that had been used in one form or another for thousands of years.

The ceremony did not remove the substantial diplomatic hurdles that still stood in the way. Though Lesseps orchestrated that day in April, he could not both micromanage the construction and maintain pressure on the British, the Ottomans, Said, and the emperor. No one could replace Lesseps in the political arena, but after a brief search, the company hired a general contractor named Alphonse Hardon. Hardon had never set foot in Egypt, but he had overseen railroad projects in Europe and had earned accolades for the construction of a number of railway stations. As general contractor for the Suez Canal, he was given wide autonomy to hire workers and to engage subcontractors who would erect dwellings, set up condensation plants, assemble dredgers, and obtain stone and lime for the jetty. The details on the ground and choice of materials were left to Mougel, who in turn had a staff of engineers under him. All of these men were handsomely paid, and Hardon personally was guaranteed a substantial bonus if he completed the work on time and under cost.[3]

As Hardon and Mougel scrambled to find workers, Lesseps left for Egypt after throwing a celebratory banquet for the senior members of the Canal Company in Paris. In early spring, he arrived in Alexandria and met with Said. He informed the pasha that Napoleon III wanted the canal built, and that he intended to honor the emperor's desires. Said complained that Lesseps was once again placing him in an untenable position by invoking the emperor's name without any concrete evidence that this was the emperor's policy. Said also bristled at what he rightly took to be Lesseps's impertinence. Rather than approaching the pasha as a supplicant seeking a blessing, Lesseps acted like a business partner. Said, ever polite and still averse to making a scene, said nothing. Informed that Lesseps had left for the isthmus to begin work, Said announced that he had not yet granted permission. Lesseps, anticipating that there would be trouble, claimed that the activity at Port Said was only a continuation of the preparations that Said had authorized

long ago. The English consul, infuriated that Lesseps had circum-
vented the will of the viceroy, wrote angrily that Said was "a man without moral courage or consistency, who cannot be relied on for a day." In
meetings with both French and English diplomats, Said refused either
to disavow or to sanction what Lesseps was doing. And so the dance
continued.

But this time, the opponents pressed harder. Within the Egyptian
ruling class, there were some who sided with the English. One of
these was the minister of foreign affairs, Sharif Pasha, who instructed
Lesseps that the prior permission to conduct preliminary works could
not be used as a pretext to commence building the actual canal. He
ordered Lesseps to cease and circulated this order to the consular
community in Egypt. Lesseps politely refused. Until such a directive
came directly from the viceroy, he planned to call all bluffs. He wrote
to Said warning that "the adversaries of the Suez Canal are the enemies of Egypt and of the viceregal dynasty." He confessed his surprise
that Said "had given in to the pressure of these enemies and committed such a flagrant injustice" in allowing his minister to issue a statement that violated earlier promises to the Canal Company. "Your
Highness," Lesseps scolded, "is not at liberty to dismiss the sacred
engagements, contracted in full view of the civilized world, and I am
not free to delay . . . their fulfillment. The company is irrevocably constituted by virtue of inviolable decrees, and it must be allowed to exercise freely . . . the rights that it has been granted. To bring to an abrupt
halt, after seven months of existence, its legitimate exercise of these
rights would be to compromise your responsibilities in the most grave
manner."[4]

Said felt squeezed. He was also receiving letters from Constantinople advising him to think carefully before allowing the canal to proceed
any further. The vizier, Ali Pasha, chided Said for not consulting with
the sultan about a matter of such clear importance to the Ottoman
Empire, though he stopped short of invoking the sultan's authority and
ordering a halt to the company's activities. Feeling pressed from all
sides, Said complained to the British consul that it was all fine and well
for Palmerston (who was once again prime minister) to advise that the
Egyptian government intercede, but not if Britain was simply using
Egypt as a pawn. Said resented being put in the middle. "France wants
a canal, and England does not," he said, reasonably enough. No matter
what he did, one power would be offended, and unlike the Ottoman

ministers in Constantinople, he did not have the luxury of inaction. It was easy for the English and the grand vizier to pressure him, but he had authorized the creation of the Canal Company, and he alone would suffer if he tried to break his commitments.[5]

Meanwhile, work was continuing, though hardly at a breakneck pace. There were still fewer than two hundred people living at Port Said, and only the barest outlines of the town and the jetty had taken shape. A temporary wooden jetty had been built to accommodate larger steamers that carried water, building materials, and stone from the quarries of Mex, outside of Alexandria. The company started to hire native workers to assist in the labor. But the intransigence of London and Constantinople was having an effect. There was only so much that could be done with the limited supply of workers, and until Said allowed the company to recruit large numbers of Egyptians, progress would be minimal. The freshwater canal had to be dug before substantial work could begin in the isthmus itself, and no matter how broadly the company stretched the definition of "preliminary," the freshwater canal would never fit. Aware that this stasis suited only the canal's opponents, Lesseps lobbied Said with a steady stream of legalistic letters. He also escorted Said's son Tousoun through Paris in the summer of 1859, as an act of friendship and as a signal that his relationship with Said remained strong.

But though the viceroy trusted Lesseps to take good care of his son, he was slowly succumbing to the arguments against the canal. In October, Said officially instructed the Canal Company to cease and desist. This time, the order was serious. The viceroy forbade further shipments of supplies from Alexandria or Damietta. Without these, the company would be forced to evacuate Port Said within a matter of weeks. There was only one recourse. Lesseps appealed directly to the emperor.

Napoleon was aware of the controversy. Lesseps had written in August claiming that the arguments against the canal were scurrilous and stemmed only from Great Britain's animosity toward France, but Napoleon had once again declined to become involved in the dispute. Now, however, he agreed to meet with Lesseps. The emperor was at the height of his power. He had tightened his hold on France after the Orsini plot, and French armies had just scored a decisive victory over the Austrians at the battle of Magenta (which promptly became the name of an avenue in Haussmann's ever-changing Paris). More popular than ever, Napoleon was less concerned about upsetting Britain and more determined to extend French influence throughout the world. That led to a few inane policies. Convinced of French greatness, he sent armies to Mexico to support the Archduke Maximilian's effort to install a Hapsburg monarchy in the land of the Aztecs. But Napoleon's expansive confidence also led him to rescue a Suez Canal project that was wildly popular among the same classes in France that formed the bulk of his support.

The emperor granted Lesseps an audience toward the end of October. There is no record of what role, if any, Eugénie played in arranging the meeting, and only Lesseps left a written account of what transpired. The meeting took place at St. Cloud Château, on the outskirts of Paris, and the emperor skipped over the formalities and went right to the point. "How is it," he asked the entrepreneur, "that so much of the world is against your enterprise?"

"Sire," Lesseps responded cleverly, "it is because the whole world thinks that Your Majesty does not support it."

The emperor considered these words carefully, while absentmindedly twirling his mustache, as he often did when he was confronted with a question that required some thought. "Ah well, rest assured," he told a relieved Lesseps. "You can count on my protection."

The meeting continued for several more minutes, while Lesseps wrangled from the emperor an agreement to replace the French consul in Egypt, who had been less than helpful to the Canal Company. After hurrying back to the company's office in central Paris, Lesseps publicized what the emperor had told him. Napoleon himself did not issue a statement confirming what he had said, but his representatives in London, Constantinople, and Alexandria demurred when furious English diplomats demanded an official denial of what Lesseps was claiming. It soon became clear that Lesseps was not lying and that the emperor had indeed blessed the enterprise. Having secured this vital backing, Lesseps proceeded to ignore Said's direct orders to halt, and Said, recognizing that the emperor himself stood behind the company, reversed course and once again became a champion of the canal.[6]

To placate the Ottomans, Lesseps returned to Constantinople at the end of 1859, to dine with Sir Henry Bulwer, the British ambassador, who had succeeded the aging Stratford Canning de Redcliffe. The grand vizier had also been replaced, and there were rumors that the sultan planned to revoke the hereditary privileges previously granted to

Said and the family of Muhammad Ali. The Ottoman ministers, not to mention the sultan, were piqued that Said had ignored their strong suggestions that he not allow the Canal Company to proceed, and they communicated their displeasure to Lesseps. But the veiled threats were just that. No one in a position of power at the Porte seriously intended to revoke Said's privileges, because no one was certain that such a decree, if issued, would be obeyed. The viceroy of Egypt may have served at the pleasure of the sultan, but unless the sultan was willing to dispatch an army, he could not dramatically alter the status quo. As long as Said was alive, he and the Ottomans engaged in an elaborate charade: they pretended to rule over him as long as he continued to speak as if that were the case, and he continued to speak as if that were the case as long as they did not act as if it were.

Far from conceding the war, in the spring of 1860 Palmerston and the canal's opponents began the next battle. New arguments were made about the dangers of the Red Sea, and articles started to appear questioning whether ships could even navigate through its supposedly erratic and violent currents. As long as little actual work had been done, there was still some hope that the tide of informed opinion could be turned against the Canal Company, and that this would then undermine the confidence of investors. Though the chances of this were slim, the continued opposition cost little to implement, and it forced Lesseps and the company agents to waste precious time and resources defending themselves. Palmerston also encouraged British diplomats to maintain the drumbeat against the canal. On the initiative of the new consul in Egypt, a "fact-finding" mission was dispatched to the isthmus. Lesseps had been claiming that the work was well advanced. Apparently, that was not the case.

According to the report filed by the consul, in the entire isthmus, the company employed a total of 210 Europeans and 544 Egyptians, not including staff in offices in Cairo and Alexandria. The company's agents had marked out eleven stations along the proposed route, with anywhere from three to a hundred Europeans camped at each point. But most stations had fewer than half a dozen Europeans and a dozen Arab workers living in tents. The major settlements, at Port Said, Lake Timsah, and Kantara, consisted of a few dozen adobe huts each, with a general store, lime and brick kilns, wells, a few steam engines, and dilapidated old dredgers. The outpost of Gebel Genifa, twenty-eight miles northwest of Suez, had been praised by the company's propaganda as a thriving community. Yet it consisted mostly of stakes in the sand marking out future streets that had already been given names such as Rue de Lesseps and Rue de Mougel. The soil consisted of sand, clay, and some lime. Even with fresh water, it would be useless for agriculture, and the well water was so saline that it was essentially undrinkable.

At Port Said, the busiest of the stations, there were two dredging machines for work on the small access canal and a total of eleven wooden huts. While the residents spoke of plans for engines to condense steam into drinking water and boasted of the stone that was supposed to arrive from Attaka and Alexandria, none of this had yet happened, and morale was low. One engineer complained that the company was disorganized. "They give me men, and no pick-axes; another time they give me pick-axes without men; and once or twice, when they have succeeded in giving me both at the same time, they have found it impossible to send provisions." Drunken fights between the residents were common, though at least they could get their broken bones set and their cuts stitched at a makeshift hospital, staffed with European doctors hired to tend to workers who would almost certainly be injured when construction of the jetty began.

Details of the report were leaked to the press. *The Times* of London wrote, "A show of carrying on works in the Suez Desert is still kept up, but every one is agreed that it is nothing more than a mere pretense, and that not the slightest real progress is made." The paper claimed that, though the project could be completed with enough money, the sum required would be many times the amount currently budgeted. As things stood, *The Times* concluded, the company was about to spend eight million pounds (the equivalent of two hundred million francs) "digging holes in the sand" that would soon fill up again, leaving the isthmus as impassable as ever.[7]

Lesseps painted a very different picture of the scene. In June, he left Alexandria for the canal zone, and though he was president of a company and not a country, his tour had all the trappings of a state visit. Accompanying him were the chief engineers of Mougel's staff, a number of French naval officers, a priest, and a Muslim imam. After surveying the quarries at Mex, the party made the difficult trip east to Port Said. The route was by this time well established, but it was still an arduous journey. No large ships could actually dock at Port Said itself. The group had to take a train that terminated about forty miles from Damietta. Then they disembarked and went by mule to the village of Mansourah, where they got on barges that took an entire night to navigate the eddies to Damietta. From there, they went by lake boat across Manzala to Port Said. The whole trip took three days.

The night they arrived, the priest celebrated mass, and then baptized the first child born in Port Said. Lesseps was honored to be the godfather, and the child gained two additional names, Ferdinand and Said. In return, Lesseps gave the child two shares in the company. Mass was followed by a banquet hosted by the company for the entire community.

Lesseps was delighted by the sight of the new fifty-foot-high lighthouse, and he wrote glowingly about the town. It was, he claimed, filled with new stores, including a butcher shop, a bakery, a canteen for supplies, and a restaurant. Other stores sold shoes and clothing. Food was dispensed at reasonable prices, and a bottle of red wine from France went for about half a franc. Though there were still fewer than five hundred workers, there was already a nascent system of payment. Employees of the company were given color-coded chits corresponding to their salaries, and they used these to purchase whatever they required. Soon, arrangements would be made for licensed Egyptian vendors to sell the company workers everything from fish from Lake Manzala to fruit, vegetables, and eggs, and deals were being finalized to establish a distribution chain, with company-owned warehouses in Damietta on the coast and at Bulaq on the Nile. Materials and dry goods could then be transported as needed to the eleven stations along the canal route.

Slowly, the company was acquiring the powers of a small state. In America and in Europe, both railroad and mining corporations perfected the art of the company town. Rather than being paid a direct salary, workers were issued credits for use at company stores where they could buy sundries and necessary amenities. In time, the Suez Canal Company became a bureaucracy that ran not just the canal, but also the lives of hundreds of thousands of people in the canal zone. In the spring of 1860, Lesseps could only hope that such a day would arrive, but already there were signs of what was to come.

Materials were being assembled at Port Said. Rail tracks and digging tools were piled up near the makeshift jetty, and an access rail line was being laid next to the canal to facilitate the movement of dredgers, men, and equipment. It would not be operational for many months, however, and Lesseps's party toured the canal zone between Port Said and Lake Timsah by camel and mule. They inspected the stations along the way, and stopped at Kantara and at the future site of Ismailia (which was then called Tousounville, after Said's son) before returning to Cairo along the proposed route of the freshwater canal.[8]

Though radically different in tone, the two accounts are reconcilable. The report filed by the British consul and the version left by Lesseps agree in most respects. The difference was in how the facts were interpreted. The British filtered the data through a lens of hostility. They wanted the canal to fail, and so, instead of lauding the bare-bones settlement, they pointed to the primitiveness of Port Said as evidence that the project consisted mostly of propaganda. Where Lesseps saw the beginning of a glorious future, the British observer saw a pitiful group of single-story dwellings dwarfed by a rickety light-house and a pier that could barely contain the mild tides. Where Lesseps discerned the outline of a railway to supply the rest of the isthmus, the British noticed piles of rusting metal. And whereas the small community would have cleaned up and made everything look its best in anticipation of Lesseps's inspection, the British observer arrived anonymously and was able to observe the colony in its less polished state.

No matter how the picture was interpreted, as of the summer of 1860, the company was at best a chrysalis of what would be. Though Port Said had grown, there was no evidence of an actual canal, and, however unfair, that fact alone provided ammunition for the project's determined adversaries. When Said was presented with the bill for the shares he had agreed to purchase, Lord Palmerston used these reports to assail Lesseps and the company for having misled investors.

It was no secret that the viceroy intended to pay for the shares that had not been bought in Britain, Austria, Russia, and the United States. But that did not stop Palmerston from making an issue of it. Said was asked to pay an installment of fifteen million francs on the 177,642 shares that he owned. Compared with the revenue taken in annually by the Egyptian government, that was a substantial sum, and though Said could cover the first payment from the reserves in his treasury, beginning in 1862 he had to borrow money from European bankers in order to meet his obligations. Palmerston accused Lesseps of taking advantage of the viceroy. "A great many persons in France— small people—have been induced to take small shares," he declared in

Parliament, "under the notion that the concern would be a profitable matter. . . . The progress of the works in Egypt, however, has been such as to show that, if not impracticable, it will require an expenditure of money, time, and labour quite beyond the reach of any private company." But, Palmerston continued, "the unfortunate Pasha" had been "hoodwinked" by Lesseps into thinking that the expenditure would be good for Egypt. Though he did not doubt that it would be good for Lesseps, he feared that it would lead to Said's ruin. It was not, Palmerston claimed, simply a private matter between a French company and a foreign investor. If Said was forced to go into debt to pay for the shares, and if he contracted loans with French bankers and then defaulted, then these bankers, in league with the French government, might start seizing property, or even take control of Egypt's finances, in order to collect their money.[9]

Palmerston was wrong yet prescient. The ruler of Egypt did go into debt. He did default, and European bankers did take control of Egyptian finances, which paved the way for the collapse of the Egyptian state and the occupation of the country by a European power. But the ruler was Said's successor, Ismail; the bankers were English; and the European power was Great Britain. Though the canal contributed to the bankruptcy of Egypt, as Palmerston feared, it ultimately enriched Great Britain and enhanced British power, as Palmerston would have liked.

Said graciously paid the first installment, but the extra money did not provide any substantial boost to the work. The situation on the ground was certainly better than the English claimed. After all, until a few years before, the entire region had been inhabited only by a few fishermen trawling the swamps of Lake Manzala and by Bedouin tribes who occasionally camped along the isthmus. That the company towns at Timsah and Port Said existed at all was an accomplishment, and even under the best management, it would have been difficult to set up supply lines any more quickly.

But it soon became clear that the company was not graced with the best management, or with the best laborers. Hardon and Mougel were not working well together. In part, this was an issue of personality, but there were structural problems that made the situation worse. Never having been to Egypt, Hardon had agreed to serve as general contractor without a full appreciation of the immensity of the undertaking. Faced with the actual problems of a waterless desert region, and confronted with the reality that supplies and workers would have to be obtained elsewhere and then transported to the isthmus at considerable cost, Hardon was flummoxed. It was not that the work wasn't getting done, but it was taking far longer than planned.

Whereas Hardon suffered from lack of experience in Egypt, Mougel may have had too much. He gathered a superb staff, but having spent decades working on plans for the canal, he had a hard time integrating new information. He thought he understood what needed to be done better than anyone else, and he might have. But as more people with expertise divided the work into component parts and began to examine the soil and the precise route and what was required in each place, Mougel resisted change. He was wedded to the initial blueprints, and as the site engineers started making adjustments, he was slow to alter the master plan. That meant delays in ordering new supplies, in recalculating costs, and in revising the route.

For instance, there was the challenge of excavating Lake Manzala. The first section of the canal was supposed to run along its eastern shore, which meant that a channel had to be dug out of the marshy, sulfurous lake-bed. From his experience as the chief engineer on the Nile dam, Mougel planned to use a few dredgers to carve a channel through the lake, but the dredgers he obtained didn't work properly. There were problems mooring them along the shore, and even when they scooped out soil from the lake bed, the currents from the lake carried silt that rapidly filled in whatever holes had been dug. The dredgers could go for hours and have nothing to show for it.

But the fishermen of Lake Manzala knew how to manipulate the lake. They had been doing so in small ways for thousands of years. The company hired these fishermen, who stood knee-deep in the water and scooped up as much as their arms could hold. They then rolled the mud, which stank of sulfur, into balls, squeezed out the water on their chests, and let these balls bake in the sun until they hardened. The hardened round bricks were used to form a break wall that rose above the lake's surface. Only then could the excavation of the channel proceed. At each stage of the work, improvisation like this was required. The method for building an embankment near Port Said didn't work farther south, near Lake Balah, because the soil was completely different. Manzala soil hardened, but the gypsum of Lake Balah cracked, which meant that material for the embankment had to be trucked in from somewhere else.[10]

In fairness to Mougel and Hardon, they had been put in an untenable position. They had been instructed to build a canal and were expected to solve whatever problems arose. But the fault lay as much with Lesseps and the company as it did with them. There was no way that they could not have made mistakes, and the company's expectations were unreasonable. But by the middle of 1861, Lesseps decided that both men had to be replaced. That meant firing the two senior officials involved in the construction. No company undertakes a decision like this lightly. There was a very real chance that the removal of Hardon and Mougel would disrupt the timetable, jeopardize relationships with subcontractors and suppliers, and cause an uproar among the shareholders. But Lesseps was certain that neither man could complete the project, and whatever the risks of removing them, he believed that keeping them would be worse.

Hardon remained until his contract expired at the end of 1862, and Mougel was slowly eased out. Their successors fared much better, in part because they were able to learn from the early false starts. In January 1861, François-Philippe Voisin was appointed engineer-in-chief of the canal works; Mougel remained director general for another year. A graduate of the Polytechnic, Voisin had established his reputation working on the port of Boulogne and then on hydraulic projects in the Pyrenees. Soon known as Voisin Bey, he was forty years old, energetic, and, as it would turn out, perfectly suited for the task. Lesseps was delighted. He announced that he had "unlimited confidence" in Voisin and praised his "excellent judgment and good nature." The canal had found its engineer.[11]

But one other problem remained unresolved: workers. For the first two years, the company relied on voluntary laborers. Said had promised to provide the project with labor, but Lesseps was wary. The corvée would give the company tens of thousands of workers, courtesy of the Egyptian government. But even though these workers would be paid, they were still forced labor—not exactly slaves, but not exactly free. Popular opinion in France and England abhorred slavery. The United States was beginning a civil war because of slavery, and even the Russian tsar was coming under pressure to free the serfs. Lesseps wanted to avoid being portrayed as a man who used slave labor to build his ode to civilization.

For the first year, the corvée was not an option. Until the end of 1859, Said objected to what the company was doing and would not have authorized the corvée even if Lesseps had wanted it. Said's position changed in 1860, but Lesseps worried that "the theme of the corvée would become the pivot of attacks launched against the canal by its adversaries." Instead, he tried to find workers without the intervention of the Egyptian government. He met with Bedouin shaikhs on each side of the isthmus and asked them to supply some men. They entertained him in their tents, drank coffee with him, and smoked pipes, but they did not oblige. Though the company was able to hire several hundred Egyptians and a few Syrians and Palestinians, that was a pittance compared with what was needed. Villagers were understandably wary of foreigners promising wages for work to be conducted many days from their homes. Desperate, the company recruited as far afield as Cuba, California, Australia, and even China. None of these yielded more than a handful of men.[12]

The Suez Canal could never be created with a few hundred workers. In fact, the company did not even have enough men to dig the freshwater canal from the Nile. Though he was aware that the corvée might become a public-relations disaster, Lesseps had exhausted all other options. After the summer of 1861, thousands of peasants started to arrive in the canal zone. They were not there by choice.

# CHAPTER THIRTEEN

# THE CORVÉE

THE FRENCH EXPRESSION *"Quelle corvée!"* translates as "What a drag!" The actual corvée in nineteenth-century Egypt was that and more. Hardly unique to Egypt, the corvée has been used in countless countries for as long as human history has been recorded. Most of the wonders of the ancient world were built through forced labor. Even the temples of the Greeks were constructed by laborers who were not given the option of refusing to haul the tons of stone and thousands of bricks needed to construct a Parthenon. The Great Pyramids at Giza, the Great Wall of China, the temple of Angkor Wat in Cambodia—all were the products of hundreds of thousands of peasants who were corralled by armies of the state, taken to the sites, and put to work. Whether they survived was not a primary concern.

One of the byproducts of the English revolutions of the seventeenth century, the American and French revolutions of the eighteenth century, and the emergence of liberal ideas in the nineteenth century was that slavery and forced labor came under attack. Human bondage was difficult to reconcile with the principles of liberty and freedom. The British officially outlawed the slave trade in 1807. In France, the Society for the Abolition of Slavery agitated for decades before it was finally victorious, during the brief Second Republic interregnum, in 1848. Neither of these laws actually ended slavery in the colonies held by Britain and France, and it would take continuous effort by antislavery advocates to eradicate the human trade in Africa, Asia, and even Latin America. But by midcentury, it was difficult to find anyone in a position of influence in the West defending slavery, at least outside of the Southern part of the United States.

The Egyptian corvée existed in a gray zone. The peasants pressed into service were not considered property of the state or of private individuals. But the system did impose a form of temporary servitude. The workers could not come and go as they pleased. They were recruited under the threat of violence and forced to work under the threat of violence. Typically, the corvée was used for irrigation projects on the Nile. These took several months to construct, but it was simple to assemble large numbers of peasants from nearby villages. The Nile Delta, to the north of Cairo, was the most densely populated region of the country outside of the cities, and most of the dikes, dams, embankments, and sluices that needed maintenance and dredging were within a few days' journey from even the most remote delta village. The system was further softened by an unspoken understanding that irrigation did not just serve the interests of the few: agriculture was vital for ruler and peasant alike.

Another mitigating factor was the culture itself. Slavery and the rights of man were the subject of fierce debates in Western Europe at the time, but the same was not true in Egypt or anywhere else in the Near East and Africa. The slave trade across the Sahara continued as it had for centuries, and though African and Arab slavery may have lacked the brutality of human bondage in the United States and the Caribbean, it was still a mercenary business that treated individuals as property to be bought, sold, and used. One of the more popular destinations for Western tourists in Cairo in these years was the slave market, where men, women, and children, often from Nubia and sub-Saharan Africa, were displayed and auctioned. Compared with the slave trade, the corvée was milquetoast.

And after thousands of years, the corvée was accepted by many Egyptians as part of life. It may not have been an appealing activity, but neither was death, and both were unavoidable. The average Egyptian fellah was far more afraid of being conscripted into the army than of serving on the corvée. Muhammad Ali and his son Ibrahim had been particularly vicious in rounding up fellahin for military service, and men had been known to cut off their own fingers (so that they could not shoot) or toes (so that they could not march) in order to avoid serving as soldiers in the pasha's army. Though the fellahin sometimes resisted the corvée, they knew that they would soon be allowed to return to their villages.[1]

At its height, the corvée raised by Said for the Suez Canal involved more than sixty thousand fellahin. The goal was to have at least twenty thousand workers, but that meant that the actual number was significantly greater. It was common for critics at the time to describe the process as twenty thousand going, twenty thousand coming, and twenty thousand actually working. That made the system seem neater than it actually was. In truth, the circle was more fluid. At any given time, there was a constant flow of recruitment, transportation, work, and departure. Though laborers were often moved in large masses, it was not a simple case of transporting tens of thousands en bloc. That would have been more complicated, and potentially more dangerous. Though fellahin rarely revolted against the corvée, it was safer for several dozen soldiers to guard several hundred fellahin on their journey from the Nile to the isthmus than for several hundred to guard ten or twenty thousand.

Corvée labor in 1861 was concentrated on the Sweet Water Canal and between Port Said and Kantara, just south of Lake Manzala and just north of Lake Balah. Once in the canal zone, the fellahin required few soldiers to guard them. At least sixty miles of arid desert separated them from their homes, and Bedouin tribes that roamed the desert east of the Nile and west of Suez were not friendly. For most of the fellahin, prudence dictated serving for the few months required and digging where they were told. They were fed, and the money they would be paid once their tour of duty was finished might be useful at home. The real risk was to agriculture along the Nile. Rulers traditionally had to balance the need for the corvée with the demands of tending the fields. The able-bodied men of the corvée also grew and harvested the crops. At the time of the Suez Canal, many of these were cash crops mandated by the viceroy, and no allowances were made for cotton lost because there weren't enough men to harvest it. The fellahin were somehow expected to do both.

Adding to the difficulty were the work conditions. The Canal Company sometimes provided rudimentary shelter in the form of tents or blankets. But the fellahin were frequently left to sleep in the open near where they were digging. During the summer, that was not a problem: the desert was pleasantly cool at night. But at other times of the year, the open sands were frigid, and without proper shelter workers could succumb to hypothermia. Though Europeans assumed that the fellahin understood how to live in the desert, that was simply not true. Most of them lived in Nile villages that had a constant supply of water, abundant date and palm trees, and were sheltered from the winds by the cliffs that lined the river. The fellahin were no more adept at desert living than the average inhabitant of Paris was adept at winemaking.

There is no consistent evidence about how much violence the workers were subjected to, but the Canal Company's public pronouncements that they were well treated only made sense in relative terms. Compared with the way soldiers in Muhammad Ali's army were disciplined, or with the way slaves were handled, the fellahin did not fare badly. Compared with daily life in their villages, the corvée was a poor substitute.

The outrage that greeted the Canal Company's use of the corvée was one part moral indignation and one part double standard. British railroad projects in Egypt had used forced labor, and yet few in England objected. Stephenson's Cairo-to-Suez rail line was completed largely by corvée; had he not died in 1859, he might not have escaped the same criticism that befell Lesseps. The other double standard was that British and French factory-workers and miners were not much better treated than the fellahin, and the British habit of impressing sailors into the navy early in the nineteenth century was hardly more civil than Muhammad Ali's methods for raising an army. Even though European critics professed to be shocked at the practices prevalent in Egypt, they were less attentive to the motes in their own eyes.

With thousands of fellahin digging by hand, the Sweet Water Canal quickly reached the isthmus, and in February 1862 it arrived at Lake Timsah. A branch was extended north toward the El-Guisr plateau, and a wider branch turned south toward the port of Suez. The Sweet Water Canal had involved the excavation of 1.1 million cubic meters, which was barely more than 1 percent of the total amount that needed to be removed for the actual canal linking the two seas. As a result of the completion of the Sweet Water Canal, the isthmus finally had a consistent supply of fresh water, and that made it possible to employ far larger numbers of workers. The Sweet Water Canal was also used to transport materials, food, and tools to the work zone between Kantara and Lake Timsah, where another contingent of the corvée was carving out the narrow *rigole de service* (service canal). Lesseps dearly wanted the *rigole* finished by the late spring, but he fell short. It took some time to develop an effective system of using the thousands of fellahin and supplying them with food and water, and at the same time, Mougel was being phased out and Voisin was assuming more responsibility.

The digging went smoothly, but completion of the *rigole* took many more months.

Without the fellahin, the construction of the Suez Canal would not have been possible. Lesseps and Said were not about to dig through the desert themselves, nor were those twenty-one thousand French shareholders. Yet, vital though the fellahin were, in contemporary accounts of the canal's creation and in many books since, they are discussed en masse, as if they had one corporate identity and no individuality. This was not merely European prejudice. The Turkish-Egyptian ruling classes were just as apt to dismiss the fellahin as ignorant, illiterate peasants with minds as empty as the Sahara. The peasants of the Nile had developed a reputation for passivity that some ascribed to the climate, others to history, and it is true that Egypt's past is unusually free of peasant uprisings. But, although the fellahin may have been less aggressive than comparable populations in other parts of the world, there was more to their existence than waking, eating, working, and sleeping. That should be self-evident, yet judging from the way they have often been depicted, it is not.

Crops and floods were the central concerns of the fellahin. That had been true for thousands of years. Agriculture depended on the rising and falling of the Nile, but the Nile was irregular. Though it always flooded, it rarely did so in a consistent pattern. That could wreak havoc on the towns along its banks. To offset the destructive potential of these floods, Egyptians had spent centuries trying to master the Nile, but the contest had so far been a draw. Some years, in some places, the waters flowed through the eddies and channels around the villages and into the fields. Other times, in other places, the river crested quickly, flooded homes, and inundated crops. But because there was always water and the soil was fertile, food was rarely a problem. Even if tax revenues could be erratic, Egypt had been known as the breadbasket of the Mediterranean since the time of the Romans, and it had never failed to live up to that reputation.

It was common for visiting Europeans in this period to describe life in rural Egypt as static. Some aspects of it were, yet life had evolved over the centuries. Though the realities of the Nile and crops may have been similar for the slaves who toiled for the pharaohs and for the fellahin who worked the lands of Muhammad Ali, daily life had developed different cadences. Islam arrived in Egypt in the seventh century, and not until the tenth century were Muslims a majority of the population. A core of Coptic Christians remained in Egypt, constituting a significant minority along a stretch of the Nile north of Luxor and in Cairo itself, where a handful of churches predated even the oldest of the grand mosques. Islam itself was not an unchanging monolith. By the time of the Ottoman conquest in the sixteenth century, the country had been ruled by a succession of Arab, Kurdish, and Turkish dynasties, each of which imported its own variant of Islam. In the countryside, meanwhile, the fellahin created a folk religion that had little to do with the faith guarded by the theologians of al-Azhar University in Cairo and fought over by rival Muslim rulers.

Much of the time, Egyptian villages governed themselves. Until the reforms of Muhammad Ali, the fellahin had little contact with the central government. Life was arranged around a lattice of relationships, and the only officials were the imam of the local mosque, the government-appointed tax farmer, and the shaikh of the village Sufi order. And of these, the most consequential was the Sufi shaikh. More than the prayer leader of the mosque, more than any official of the government, after the sixteenth century the Sufi orders were the centripetal force that kept society from spinning out of control.

Sufism emerged gradually in the Muslim world. Beginning in the eighth century, Sufi mystics stressed a direct, individual connection with God. In time, these mystics drew followers, and these followers began to codify the lessons learned from their masters. Eventually, much as monastic orders coalesced in Europe, Sufi orders formed, and were named after their putative founding teacher. As the Muslim world fragmented into numerous warring states and competing sects, Sufi orders filled the vacuum. After the Mongol invasion of the thirteenth century eviscerated Baghdad and other centers of Muslim civilization, Sufism assumed a dominant place in the lives of millions.

In Egypt especially, the Sufi orders were intimately woven into society. By the time of Napoleon's invasion, the local Sufi shaikh performed weddings and funerals, adjudicated disputes, offered counsel and advice, doled out loans, and provided spiritual guidance to the inhabitants of the village. Holidays often revolved around the birthdays of notable shaikhs. At Tanta, the largest settlement in the delta, thousands gathered every October for a week-long festival in honor of Shaikh Ahmad Badawi, who had lived in the thirteenth century and whose tomb and shrine became the geographic center of the Badawiyya order. During the celebrations, animals were slaughtered, prostitutes took a working holiday, different social classes mingled, prayers were said, and people consumed an abundance of sugar-coated nuts called *hubb al-Aziz* ("seeds of the beloved Prophet").

Badawi's shrine at Tanta was known throughout Egypt, but the country was dotted with local shrines of Sufi saints. These were often surrounded by a small complex of buildings that included not just the grave of the saint but a structure large enough for the brotherhood of the order to congregate and for living quarters for the administrators. The complexes also had lands attached to them, and the income from these fields supported the brotherhood.

One of Muhammad Ali's many reforms was to seize the land of these local brotherhoods and appropriate them for himself, much as Henry VIII took possession of the Catholic monasteries of England when he severed his connections with the Vatican. But, unlike the British monarch, Muhammad Ali did not care about theology. Though the pasha was not shy about using violence, even he had to tread delicately when he dispossessed the Sufi lodges of their land. He altered the tax system so that revenue flowed directly into his treasury, but for the most part, he did not interfere with local customs. When he made the Sufi shaikhs into paid servants of the state, he did the same for the religious scholars in Cairo. But he did not dictate what they taught or preached. A balance was reached. In return for a steady income from the state, the shaikhs supported Muhammad Ali and his immediate successors. The Sufi lodges remained as vital as ever, even though they had lost some of their autonomy.

Europeans who arrived in Egypt in these years were fascinated with the Sufis. Egyptian folk religion was a cacophonous amalgam of rites. Some were a product of centuries of Sufi practice. The goal of mystics everywhere has been to find the path to God, and different Sufi orders discovered different paths. Some used chanting and breathing; believers would recite the name of God or verses of the Koran as a form of meditation. Others danced. The most famous of the dancing Sufis followed the teachings of the great master Rumi and became known in the West as the whirling dervishes. The orders that encouraged dancing, singing, and chanting were sometimes known as the "ecstatic orders," as opposed to the "sober orders" that abhorred such activities. Most Egyptian Sufis were part of the first group. Ahmad Badawi had been known to shout continuously for days, working himself into a

trance. He also fasted, and once stood staring into the sun until his
eyes turned bright red. But whether emulating Badawi through the vis-
ceral ecstasy of chants or copying other masters by pursuing the solitary life of hermits, all Sufis strove to remove the barriers that
separated individuals from Allah.

In the Egyptian countryside, Sufism merged with local practices
and traditions that had existed long before Islam or Christianity had
appeared. Very few fellahin could read or write. The Koran was entirely
a vocal experience for them. They heard it chanted on Fridays at the
local mosque, and during the daily call to prayer. But they didn't always
understand the Arabic of the Koran, because the language they spoke
was an Egyptian dialect that differed significantly from classical Arabic. Instead, their Islam was a mishmash of fertility rites and prayers
for floods and good harvests, of weddings and funerals, and of seasonal
festivals to honor the birthdays of the Prophet and the saints.

Over time, these festivals and practices had been clothed in Sufi rituals, and their origin in pre-Islamic times was forgotten. Though saint
worship was censured in the Koran, men and women would go to the
local shrine to touch a piece of cloth that had belonged to the Sufi master or to kiss the screen in front of the tomb, in the hopes that the gesture would impart some of the saint's *baraka,* or "blessing." Shaikhs
gained influence and prestige from their ability to perform miracles and
magic. Sometimes, these miracles took the form of a barren woman
becoming pregnant, a sick child infected with worms from the Nile and
about to go blind regaining sight, or a man with several daughters being
graced with a son. But the magic could also be more prosaic. At festivals, shaikhs or younger men in the brotherhood achieved a frenzied
state after singing, chanting, and smoking a water pipe laced with
hashish. Then they would dash across a pit of hot coals, or swallow
swords and eat flames. Crowds would gather, as crowds always do
whenever anyone does something extreme, and cheer or jeer depending on the outcome.

These practices drew the attention of Europeans who wrote about
life in Egypt, and they attracted the animosity of Muslim scholars and
judges. The learned muftis of Cairo viewed the folk religion of the Nile
as un-Islamic and bathed in superstition. They had tried over the centuries to force Sufis to reform their practices and bring them in line
with the textual, legalistic Islam of al-Azhar. Judges like Jabarti, who
wrote so scathingly of the French invasion, were even more hostile to

the village shaikh. But in this tug of war, neither side had scored a decisive victory, and Egyptian society was marked by an unresolved tension between Islam as it was understood by the fellahin and Islam as it was interpreted by religious scholars.

But, however much the fellahin were shaped by an ecstatic, magical Sufism, however much they were steeped in notions of a mysterious God and the equally mysterious Koran, Islam was not especially rigid or doctrinal for them. It was simply part of the landscape, sharing time and space with daily mundaneness. Lucy Duff-Gordon, who lived in Egypt during these years, wrote perceptively about village life. "The best houses have neither paint, whitewash, plaster, bricks or windows—nor any visible roofs. They don't give one the notion of human dwelling at all at first, but soon the eye gets used to the absence of all that constitutes a house in Europe, the impression of wretchedness wears off, and one sees how picturesque they are with palm trees and tall pigeon houses and here and there a dome over a saint's tomb."

Islam was part of personal identity, but not in a self-conscious way. A fellah was a son, a man, a father, a husband, a farmer, a Muslim, or a daughter, sister, wife, and mother. God was everywhere, or nowhere. Shaikhs were loved, and sometimes feared. Rulers were distant, and best avoided. Diet was simple, but rich with dates and eggs, bread, butter, and milk. The Bedouins of the deserts may have used camels, but most of the fellahin stuck to the mule or donkey as the draft animal of choice. Women were married by their fathers and families; wives obeyed their husbands; children obeyed their parents; and everyone obeyed the government.[2]

Forced to labor on the Suez Canal, the fellahin brought their culture with them into the desert of the isthmus. They did not try to fight against the injustice of the corvée. Instead, they bided their time, tried not to work too hard, and trusted that it would soon be over and they would return to their homes. And almost to a man, they did. Though they were made to dig and shovel and haul earth for this strange new trench through the sands far away from the Nile, they were paid a few piastres a day (in the form of promissory notes redeemable only after their service was completed), fed adequately, and soon back home, worse for wear but at least not dead. True, Hardon was accused of mistreating the native workers, by failing to provide them with enough food and by allowing foremen too free a hand in doling out physical punishment. But even as the corvée came under attack, there were surprisingly few allegations of violence. The fellahin knew they would be whipped if they tried to escape and treated roughly if they shirked. All in all, it was safer to follow and obey.

In addition, partly to head off charges of abuse, the Canal Company had an extensive team of doctors trained in public health. Egypt was frequently beset by cholera breakouts, and given the vagaries of the water supply to the isthmus, cholera was an ongoing concern. But these doctors also monitored the health of the fellahin to ensure that they were receiving enough nutrition. Though some of the attentiveness to the health of the laborers was motivated by a desire to avoid bad publicity, the result was that most of the workers arrived in good health and left in good health. Lesseps was adamant that the workers be treated with respect, and he was reported to have told one of the foremen, "Treat the natives well; they are men." The moral implications of forced labor notwithstanding, the corvée used for the building of the canal was not marked by rampant brutalization of the workers.

In fact, life in the canal zone was shaped by the routine of the fellahin as much as by the needs of the company. Each encampment of workers had a representative from the Egyptian government as well as officers of the Canal Company to make sure that disputes were settled peacefully, and the fellahin's customs were honored. In Port Said, the local imam demanded that the company build a mosque and a Koran school for the children of the longer-term workers. Some of the fellahin who were brought via the corvée chose to stay, and they established their own communities next to the European-style towns designed by Lesseps and the company. On one side of the Rue de Lesseps in Tousounville, at Lake Timsah, for instance, there might be a modest single-story wooden house inhabited by two senior engineers from France. They would drink their wine in the evening, sit down to dinner, and then write reports. Nearby, a few hundred yards away, several dozen fellahin might be camped in tents, eating dates and dried chick-peas while listening to someone playing the ney pipe and someone else banging on a simple drum. And for the most part, these two worlds sat peacefully side by side, with only sand and a cultural divide separating them.[3]

Still, however civil the interactions were, forced labor was forced labor. No matter how carefully doctors inspected them and how many piastres they were promised, the fellahin were not removing millions of cubic meters of sand and dirt because they shared the ambitions of Lesseps, Said, and several thousand of the French bourgeoisie. As expected, the British seized on the corvée and used it to condemn Lesseps and the company. Before Said authorized the use of forced labor, members rose in the House of Commons to excoriate the company for even contemplating the corvée. By the middle of 1862, anti-canal sentiment was once again cresting in Britain. "It is clear," claimed Darby Griffith in Parliament, "that a great evil is being perpetrated by that company in an unblushing manner." He reminded his colleagues that the sultan himself had, in that edict of 1856, declared that "the lives, property and honour of every subject in the Ottoman dominions should be held secure." Did not the corvée violate the stated policy of the Ottoman Empire? Griffith asked. And was it not England's duty as a civilized nation to do its part to stamp out slavery throughout the world?[4]

As the British government prepared its next salvo, Said went on a summer tour of Paris and London. Several months earlier, he had inspected the progress of the canal with Lesseps as his guide, and he now hoped that the diplomatic nightmare was safely behind him. In England, he was treated politely, and in France, he took full advantage of the pleasures of food and high society. That fall, he was invited by Lesseps for another visit to the isthmus, which Lesseps hoped would become an annual event on a par with the annual shareholders' meeting as a symbolic celebration of the canal's progress. The *rigole,* the narrow access canal extending roughly fifty miles from Lake Manzala to Lake Timsah, was nearly complete. The El-Guisr plateau was the final obstacle, and it had been surmounted.

The viceroy, suffering from stomach problems, declined to make the trip. But he sent a party of notables in his stead. To mark the occasion, Lesseps had the small settlement at Tousounville brushed and shined. Ceremonial arches were erected next to the lodge that had been built to house the viceroy. A special train took the deputation from Cairo to Zagazig, and from there they followed the Sweet Water Canal until they reached Lake Timsah. The party included the grand mufti of Egypt, the Catholic bishop of the region, and the viceroy's delegate and nephew, Ismail.

Never one to forgo a symbolic gesture, Lesseps had the viceregal party assemble on the morning of November 7 on a recently constructed pavilion. The event had been carefully planned. At eleven o'clock, he raised his hand and shouted, "In the name of His Majesty Muhammad Said and by the grace of God, I command the waters of the

Mediterranean to flow into Lake Timsah!" Then workers broke through the temporary earthen sluice, and the water began to stream into a lake bed that had been dry for centuries.

Of course, a banquet followed. It had been arranged at great expense, since most of the food had to be shipped across the desert. After prayers by Catholic priests, Muslim clerics blessed Said, Lesseps, and the canal. "In the name of God the compassionate and the merciful," the mufti of Cairo intoned, "who has permitted the waters from the sky to descend to the earth . . . and who has made possible these great discoveries which make it easier for the peoples of the earth to communicate with one another." For the finale, a telegram of congratulations from Napoleon was read aloud. Lesseps glowed in the victory of the moment, relieved that, after years of uncertainty, the canal was taking physical form. He wrote a long letter to his brother-in-law heralding this latest success.[5] He did not realize how short-lived it was to be.

# CHAPTER FOURTEEN

# THE NEW VICEROY AND HIS MINISTER

I N JANUARY 1863, Muhammad Said Pasha was only forty-one years old, but his body was breaking down. He had suffered from one ailment after another, and nothing his physicians prescribed did any good. Trips to Europe, long periods of rest, and a restricted diet only made him weaker. Lavish parties, such as the one he threw for himself on his birthday in December, no longer cheered him. In years past, he had rarely exhibited outward signs of stress. Rather than grapple with difficult issues, he looked for activities to take his mind elsewhere. But this time, he could not. People began to notice that something was wrong. He looked pale and exhausted, and he smiled less.

The physical changes were reflected in his mood. Meeting with the British ambassador to the Porte, Sir Henry Bulwer, in early January, Said was depressed and pessimistic. The Egyptian economy was booming, thanks largely to a significant surge in cotton exports. Some of that good news was offset by Said's decision to secure a public loan in order to finance both the purchase of the canal shares and numerous other public-works projects. Yet all of these undertakings promised more economic activity and a brighter future for Egypt and for the viceroy. His advisers, not the most candid group but loyal to a fault, praised him and spoke soothingly of how his name would be immortalized by the great projects he had overseen. Talking with Bulwer, however, Said confessed that it all seemed pointless. He worried that the canal would become a French fiefdom, and that it would undermine Egyptian sovereignty.

By mid-January, as his health rapidly deteriorated, courtiers began to prepare for the worst. Overtures were made to his nephew and successor, Ismail, who lived on an estate in the delta. One ambitious

French gentleman employed at the viceregal court managed to plant himself in the palace telegraph office so that he could be the first to tell Ismail of Said's death. It had long been customary in Egypt for the new ruler to reward the bearer of such news with wealth and rank. Unfortunately for the ambitious courtier, Said was slow to die. After many days of little sleep, the courtier bribed a telegraph clerk to bring him the news and then went home to bed. But when the announcement of the viceroy's passing was transmitted, the clerk decided to inform Ismail himself. In return, he was granted a title and an income, while his erstwhile sponsor came away with nothing.[1]

For different reasons, Lesseps shared Said's fear of being undermined. On the night of January 17, Said died in his sleep. Ismail, as the oldest surviving male in the family, was proclaimed the new viceroy, though only provisionally. Until the sultan ratified the line of succession, Ismail could not officially claim the title, and he left for Constantinople three weeks later to secure the sultan's approval. But within days after his uncle's death, he set a markedly new tone for the country, one that alarmed the partisans of the Suez Canal. For all of his vacillations, Said had been a staunch advocate of the project. No matter how grim things looked, Lesseps had always been able to strengthen Said's resolve or persuade him to support the company. In a country where the will of the sovereign meant almost everything, that personal bond had been the difference between success and failure. Now, with Said gone, Lesseps and the canal had lost the only Egyptian patron they had. Regardless of the concessions that Said had granted, unless those guarantees were renewed by Ismail they would be worthless.

Ismail had been the heir apparent since 1858, after his older brother was killed when a train plunged off a bridge and into the Nile on the way from Alexandria to Cairo. In January 1863, he was thirty-two years old, almost the same age Said had been when he assumed power eight and a half years before. He was the son of Ibrahim Pasha, and he had spent time with his father in the Syrian desert during the campaign against the Ottomans in 1839 and 1840. Though hardened to realities of military life by a father who instilled in him a respect for the art of war, Ismail did not grow up to be a military man. But whereas Said had been mercurial and expansive, Ismail was deliberate and reserved. Bulwer, with a whiff of condescension, called him "parsimonious, careful, and methodical, and with abilities rather solid than brilliant." He predicted that Ismail's administration "would be orderly, safe, and respected."

Ismail was even more at home in Europe than his uncle had been. He graduated from St. Cyr, the French military academy, and he had spent months in various European capitals mingling with the aristocracy. His manners were impeccable, as was his dress, which was thoroughly European save for the occasional red tarboosh. Otherwise, he was usually found in polished leather boots, tapered pantaloons, a tailored buttoned frock coat adorned with military epaulets, trimmed hair, and a groomed beard. Stocky, he shared with Said a tendency to overeat, but he never approached his uncle in size.

During the previous years, Ismail had avoided attention. Though no one contested his right of succession, court intrigues could be unpredictable and potentially dangerous. Abbas had been murdered; Ismail's brother had been killed unexpectedly in an accident; and Ismail himself had no wish to share either fate. Instead, he avoided the court and quietly bought land. His low profile served him well. By the time he inherited Said's mantle, he was one of the largest landowners in Egypt, and his estates were so profitable that he had become one of the richest men in the country. He read the latest books on land use in Europe and throughout the world, and he hired consultants to advise him on how to maximize crop yields. He was an inventive overseer. Using his prerogatives as a member of the ruling family, he decided not only what would be grown on his estates, but how they would be farmed. He concentrated on cash crops, cotton as well as sugar cane. He sampled different strains of seed, redesigned irrigation, and experimented with nitrates, old-fashioned guano, and innovative fertilization techniques. Frugal yet willing to invest in new technologies, he was a shrewd manager of his finances—a skill that deserted him when he became Egypt's ruler.

Yet Ismail did more than tend his estates. He thought ahead, and by the time he became viceroy, he had a detailed agenda and a plan to carry it out. Like many rulers of his generation, he was both an ardent nationalist and an eager modernizer. He admired Europe and wanted to emulate it, and he loved Egypt and wanted it to be powerful. That meant following the path of his grandfather Muhammad Ali and his uncle Said. He once remarked, "Egypt must become part of Europe," and he meant it. He intended to increase the pace of reform; rationalize

Egyptian agriculture; streamline and centralize the government and its bureaucracy; improve and strengthen the military; and enhance Egypt's reputation as a country of ancient wonders and modern marvels.

That entailed spending more on trains, telegraphs, and waterways; building factories; reforming the judicial system; and transforming Cairo from a relic of past greatness into a metropolis filled with gardens, museums, palaces, and parks on a par with London, Paris, and Vienna. It also meant increased military spending, in order to equip the army with the latest weapons being produced by the Krupp munitions factory in Germany and with French and British armaments. To train soldiers how to use the new weapons, Ismail authorized the construction of military academies for specialized studies in artillery and cavalry. To house the new army, he continued work done by Said on an immense barrack in Cairo, at Qasr al-Nile, and he commissioned a new palace just to the west of the walled old city. Abdin Palace, sprawling and huge, was a perfect copy of a nineteenth-century Victorian building, and since the materials were imported along with the workmen, it was even more costly. Ismail himself, reputation for parsimony notwithstanding, spared few expenses in outfitting his court. For him, his family, and his retainers, he purchased only the finest china, the most elegant carriages, and the latest baubles, from ivory-tipped walking sticks to porcelain figurines to engraved crystal trays resting on the arms of sculpted cherubs.

These ambitions required large sums of money. Ismail had two solutions: expand the production of cotton, and decrease the power of the Suez Canal Company. In 1861, worldwide cotton supply plummeted because of the Civil War in the United States. Though the Southern states were able to maintain production for the first months of the conflict, shipments became erratic when the North instituted a blockade. For textile businesses in England and the continent, the Civil War presented a problem, and Egypt became the answer. In 1860, average cotton exports from Egypt amounted to about one million pounds sterling a year. By 1862, the figure was six million, and by 1865, it reached nearly eleven million pounds. In the words of one observer, "Practically every available acre in the Nile Valley was devoted to cotton. . . . The fields were covered with white bulbs and every fellah dreamed in terms of cotton." Blessed with this sudden windfall, Said had taken a loan in 1862 to finance new projects, and Ismail followed suit. Believing that the cotton revenues would be permanent, Ismail authorized hundreds of new ventures. Had the cotton boom continued, he might have been able to pay for them.[2]

The Canal Company was hindering Egyptian development. Because of the onerous burden of paying for those 177,692 shares, the government had tithed a large portion of its future revenue to the company. In time, the completed canal might generate substantial income for the government, but that lay years in the future. Ismail could not wait a decade or more to move the country in the direction he wanted it to go. International politics would not halt while Egypt tried to close the gap, and for every year Egypt dallied, it fell further behind Europe. Even with the surge of revenue from cotton, Ismail would not be able to fund his ambitions and pay for the shares unless he went deeply into debt. Though he did not mind the idea of loans, he feared the loss of autonomy. His predecessor's commitment to the Canal Company hamstrung the Egyptian government. Ismail's solution was to undo the harm by weakening the company and humbling Lesseps.

That did not mean abandoning the canal itself. The work had proceeded too far, and with Napoleon apparently in favor of its completion, Ismail did not want to incur the hostility of France. And he was not opposed to the idea of a Suez Canal, only to a canal controlled by foreign interests. The trick was to supplant Lesseps without alienating France or undermining the canal's completion. Not an easy task, but one that Ismail had obviously mulled in private for some time, because, within hours of the succession, he already knew what he was going to do. On January 20, 1863, he invited the consuls of Europe and America to the Citadel in Cairo. It was the same spot where his grandfather had slaughtered the Mamelukes. Ismail had no such plan for the consuls, but what he did altered the political landscape almost as dramatically.

Addressing the curious assembly, he was self-effacing. "I am profoundly aware of the magnitude of the task that God has imposed upon me in recalling my uncle to him. I hope, under the auspices of His Imperial Majesty the Sultan, my august sovereign, that I will be able to fulfill my task with dignity. I have unequivocally decided to consecrate to the country that I am called to administer all the energy which I possess. . . . I know that the basis of all good administration is order and economy in finances." For that reason, he announced that he would go beyond the bureaucratic reforms of Said and create a civil list. That would end the age-old custom of private payoffs and graft. Salaries of bureaucrats would be published, and appointments made solely on the basis of qualifications. That in itself was surprising. Even European nations that professed to honor the notion of a professional civil service rarely did so in practice, and in the United States it was not until the 1880s that civil-service exams replaced the spoils system and blatant favoritism. But the next thing Ismail said astonished the consuls. It was, he declared, his intention to abolish the corvée.

He made it clear that this was not a caprice. The system of forced labor might have worked in past eras, he said, but it now "prevented the country from developing as fully as it was capable of." In an argument that drew on the best traditions of European liberalism, he spoke of the moral injustice that the corvée represented, about how it robbed the fellahin of their God-given dignity and their rights as human beings. British officials were delighted, and the English press began to write the canal's obituary. "The work cannot be carried on in the future without enormous expense," wrote *The Standard*. "The poor stockholders in France, Egypt, and Turkey are ruined. The affair on which they have set their hearts will be barren of profits." *The Spectator* was equally convinced of the project's demise. "As forced labor is now to cease," it editorialized, "the canal ceases."

Ismail assured the diplomatic community that his target was not the canal per se. "I am more canaliste than M. De Lesseps," he told the French consul general, "but I am also of a more positive mind. I believe that no work is more grand, and that none will be more productive for Egypt. But at the present moment, its bases are uncertain and badly defined. I will affirm them, and then, surpassing my predecessor, I will push the works to their completion." His ultimate goal may have been to wrest control of the company, but as a patient, methodical man, he fought only one major battle at a time. By presenting himself as a reformer of a forced labor system that was reviled in Europe, Ismail could claim to be motivated by morality. His timing was also impeccable. Earlier that month, Abraham Lincoln had issued the Emancipation Proclamation, freeing the slaves in the Southern United States. That had been reported throughout the world, and though the issuance of the proclamation did not lead to the immediate end of slavery, it had symbolic resonance. In a similar fashion, Ismail called for an end to the corvée, though he did not immediately halt the transport of fellahin to the canal zone.

Though he justified his opposition on moral grounds, Ismail's true motives were more ambiguous. He had not taken a strong stance against the corvée before, and his personal irrigation projects had benefited from its use. Soon after announcing the abolition of the system, he did what people in positions of authority often do: he issued a "clarification." He said that, while he opposed the frivolous, abusive, and exploitative use of the corvée, he believed that it could still be levied for public works that were essential for the common good. His later actions were consonant with a strong moral animus to human bondage. He aggressively extended Egyptian influence into the Sudan for the sole purpose, he said, of eliminating slavery in East Africa. Yet that also provided an unimpeachable rationale for increasing the size of his realm and collecting more taxes.

In truth, no one has any idea what Ismail felt in his heart about the corvée. Unlike Lincoln, he had not dedicated his life to ending forced labor, and his strong opposition came as a surprise. At best, his position was a combination of expediency, opportunism, and morality. Fighting the legitimacy of the corvée would loosen the hold of the company on Egypt and gain the sympathy of many Europeans. It was a brilliant tactic, and if it also soothed the conscience of the king, all the better.[3]

In February, Ismail left for Constantinople to get the sultan's support. The trip went well, and Ismail was formally made viceroy at an investiture ceremony. Sultan Abdul Aziz was the same age as Ismail, and he had replaced his brother Abdul Mejid in mid-1861. Like Ismail, he was new to power and eager to make his mark. Breaking from tradition, Abdul Aziz began to travel abroad, and after hosting Ismail in Constantinople, he went to Egypt in April 1863. No Ottoman ruler had gone to Egypt since the conquest of the country by Selim I in the sixteenth century. Abdul Aziz treated the trip as an opportunity to remind the Egyptians that they were still part of the Ottoman Empire. Ismail looked on the visit as a chance not only to impress his overlord, but to alert him that Egypt could be a formidable competitor.

The sultan's visit was one of the most lavish affairs of Ismail's entire reign, surpassed only by the celebrations organized in 1869 for the opening of the canal. Each ruler had a specific agenda. Ismail wanted the sultan to grant him the right of succession from father to son. As it stood, the hereditary office of viceroy went to the oldest male of the family. The sultan wanted more tribute. Both anticipated protracted negotiations that would last months and probably years. These would be conducted by trusted retainers and ministers, and the two monarchs would not have discussed details. Instead, they toured the Pyramids, prayed at the tomb of Shaikh Badawi in Tanta, visited model farms, went to the antiquities museum run by Auguste Mariette, and attended one banquet after another. No record exists of what the two rulers discussed, but the Suez Canal was certainly part of the agenda. Abdul Aziz may have technically possessed the authority to determine what would be done, but that was not the point of going to Egypt. He was determined to shake things up in the empire, and he needed the revenue that Egypt provided. And though he was sultan, he was also a visitor, and he would not have been so impolite as to confront his host.

For the Suez Canal Company, these were painful months. At the end of 1862, it had seemed that nothing could halt the canal's progress. Said's death shattered the status quo. Palmerston's government was invigorated by Ismail's stance. Lesseps was demoralized. In an attempt to placate Ismail and stroke his vanity, he announced that the company would name the city by Lake Timsah "Ismailia." The viceroy may have been flattered, but his policies did not change. British officials throughout Europe and the Ottoman Empire took advantage of the situation. Once again, they raised the issue of the lands that had been granted to the company, and suggested to Ismail and to Abdul Aziz that giving the Canal Company sovereign land was humiliating, as well as illegal.

At first, Ismail listened politely to British advice, but he was offended by their presumption that they could question his legal right to do what he pleased in his own domain. He stuck to his initial strategy. He would fight the corvée rather than contest the legitimacy of the earlier concessions. He did not trust the British, and he did not believe that it was wise to attack directly a project that had the blessing of the French emperor. "Napoleon is not a man that one ought to offend, if one can avoid it," he told a minister in Constantinople after the investiture ceremony. To make his position clear, Ismail issued two new firmans that reaffirmed many of the rights of the company that had first been granted by Said. He stipulated that the company would no longer be required to pay for the extension of the Sweet Water Canal to the port of Suez. Instead, the Egyptian government would take over the cost, and the company would forfeit the lands it would have controlled along the banks for that stretch. Ismail also confirmed the obligations of his government to pay for the shares that Said had purchased, though the payments were restructured to give his treasury greater flexibility to meet the installments.

The British were incensed. Bulwer called these new concessions "stupid and traitorous measures." Palmerston's ministers contended that Ismail was overstepping his authority. Only the sultan had the right to decide the fate of the canal, and the British hinted to Ismail, as they had to Muhammad Ali and to Said, that flouting the sultan could put his viceroyalty at risk. Ismail, however, ignored these warnings and pushed ahead. He began discussions with the company about ending the corvée, and to conduct the negotiations, he deputized one of his most talented diplomats, Boghos Nubar Nubarian.[4]

Nubar was a brilliant thirty-seven-year-old Armenian, born in Smyrna (now Izmir), on the Aegean coast of Turkey, to a family of what had been known in the Ottoman Empire as "dragomen." Originally, the dragomen were simply translators, but over time they became vital intermediaries in all diplomatic and commercial interactions between the Ottomans and the Europeans. A dragoman was part diplomat, part counselor, and part translator, and because of their unique position, dragomen were quietly able to exercise enormous influence and accumulate immense wealth. Armenians and Greeks, who by virtue of ethnicity and religion were outsiders at the Ottoman court, excelled, and the official dragoman of the Porte was usually drawn from their number. Because they were Christian, Armenians and Greeks tended to be more accepted in Europe.

Though the term and the formal office of dragoman fell into disuse in the nineteenth century, the role remained. Nubar's uncle served as Muhammad Ali's translator, and he insisted that his nephew be equipped with the same skills. Nubar was sent to school in Switzerland and then went to college in Toulouse. He was adept at languages. In later years, it was rumored that he could speak as many as eleven; that may have been an exaggeration, but he did master Arabic, Turkish, French, English, Spanish, and German.

Nubar was more than a clever linguist, however. He was a cultural chameleon. Starting in the 1840s, he became the interlocutor of choice for the viceroys in delicate commercial negotiations. From an early age, he was privy to the inner circles of power. He moved as comfortably among the divans of Constantinople as he did through the salons of Paris. He dressed in the latest European fashion, though always with a maroon fez and always with the distinctive thick mustache of an Armenian gentleman. He could and did discuss world affairs with Palmerston and Stratford Canning de Redcliffe. He also could navigate through the labyrinthine financial and commercial mazes of Europe. He was a man of many countries, and he was embraced by rulers of many countries. Charming and erudite, he was one of those shadowy figures who wield power behind the scenes and without whom the gears of international relations would become hopelessly clogged.

That Nubar was able to serve successive viceroys was itself a minor miracle. From the time of the Mamelukes, Egyptian officials usually rose and fell with the ruler whom they served, and if they survived the change of regime, they were rarely appointed to positions thereafter. Nubar served not only the Europhilic Said and Ismail; he was also trusted by Abbas. It was a testament to Nubar's skill and to his ability to secure patronage that he became one of the most influential members of Ismail's court. He was nothing if not well connected. He married the daughter of an Armenian grandee in Constantinople, and her brother was close to Sultan Abdul Aziz himself. Nubar was Ismail's best asset during the years of negotiations over the status of Egypt within the empire. With perfect pitch for the nuances of Ottoman politics, Nubar knew when to use silky words and when to offer bribes, and his connections to the Porte, in addition to his extensive experience in Europe, were invaluable.

Nubar, however, was no one's servant. Though he rarely questioned his instructions, he frequently ignored them. By the end of his career, he had outlasted not just Said and Ismail but their successors as well. Ultimately, he did as much harm as good to the country he served. He helped precipitate the collapse of Ismail's regime in the 1870s, and he then tried and failed to prevent a British invasion of Egypt in 1882. But even these upheavals were not fatal to his career. He went on to become prime minister, bridling under British control and the sycophancy of the Egyptian court, before succumbing at last to advancing age. His fifty-year career spanned the last half of the nineteenth century, and, determined that his legacy and vision be understood, he wrote his memoirs in the final years of his life. It was the rich, dense volume of a proud, bitter man, chronicling the rise and decline of the heirs of Muhammad Ali, written in elegant French, full of anger toward those, like Said and Ismail, who Nubar believed had led Egypt to penury and humiliation.

Nubar remained an Egyptian patriot. He worked tirelessly to construct a country that could withstand the cultural, financial, and political pressure of Europe. His idol, not surprisingly, was Muhammad Ali.

He also respected Ismail's father and Muhammad Ali's son, Ibrahim. In Nubar's eyes, they were men of great wisdom who had changed everything in the Near East, and part of their greatness lay in their recognition of the strengths of Europe. Like them, he hoped to take the best of European culture, adapt it to Egypt, and use it to fulfill Muhammad Ali's dream of an "Arab Empire." But Nubar believed that he was more capable of serving the interests of Egypt than any of Muhammad Ali's children and grandchildren. If orders from Ismail would, in Nubar's opinion, jeopardize Egyptian autonomy, he disregarded them. That should have doomed his career, and perhaps his life, but he made himself irreplaceable. No one had his connections within diplomatic circles, and during the first years of Ismail's reign, he was indispensable not just in the struggle with the Canal Company but in negotiations with the Porte to transform Ismail from the sultan's viceroy into an autonomous ruler only nominally under the Ottoman umbrella.

Nubar had three causes: the end of the corvée, the transformation of the Egyptian judicial system, and his own enrichment. He achieved all three by the mid-1870s. He was paid substantial sums by Ismail, and he also obtained payments from the enemies of Lesseps in the 1860s and from bankers and businessmen interested in Egyptian opportunities in the 1870s. His desire for wealth is not difficult to fathom. As an Armenian in an Arab and Turkish world, he saw wealth and the good will of the ruler as the only paths to power. His interest in reforming the judicial system was a reaction against the abuses of the capitulations that exempted Europeans from Ottoman courts. European entrepreneurs who did business in Egypt could defraud Egyptians, knowing full well that their own consuls would never rule against them in the consular courts.

But Nubar's reasons for fighting the injustice of the corvée are harder to fathom. Like Ismail, he seems to have been animated by a combination of morality and expediency. In his memoirs, he described the depredations that the fellahin faced when working on the canal. Subject to beatings by indifferent overseers, they also suffered from the knowledge that back home, their crops were left unattended and might not be salvageable, which meant more debt in order to pay for food for the rest of the year. "The fellahs," he wrote to his wife in 1863, "work without pay, without food. What men! My God! They are separated for more than fifty days from their wives and children. I would like all of those who speak of how well they are treated to be treated themselves the same way."

More than Ismail, Nubar may have genuinely empathized with the plight of the Egyptian fellahin. He had absorbed European liberalism while living in Western Europe, and forced labor was incompatible with that value system. He viewed the condition of the Egyptian peasantry with a mixture of disgust and compassion. Seeing how they were treated was a continual reminder of the wide gap between Egypt and Western Europe. Of course, had Nubar taken a more critical look at Europe, he would have noticed that the lot of the Welsh coal miner or the textile worker of Lille was not much better. What he noticed instead was European arrogance, which he hated. In letters to his wife, he raged at how they treated him and other "Orientals" as children in need of instruction. His ancestors had been kings and patriarchs while the French and English were still living in mud hovels! But he respected the material progress of Europe, and wanted the same for Egypt.

Because he believed that the evolution of Western Europe demonstrated that a free labor force was a necessary prerequisite for economic growth, he was certain that Egypt could not advance economically as long as it was held back by the corvée. The abolition of the corvée would yield both material and moral benefits. As Nubar wrote in later years, "Material progress leads directly to moral progress." Given his love-hate relationship with Europe, his combustible mix of avarice and pride, his intelligence and grace, Nubar was a dangerous adversary, and as of 1863, he sought to ruin Lesseps and destroy the Suez Canal Company.[5]

The canal was still in the early stages of its construction, and many technical issues remained unresolved. As of New Year's Day 1863, there had been little doubt that it would be finished, eventually. Then Said died, and Ismail and Nubar upset the delicate balance that Lesseps had achieved after years of effort. As it turned out, they were the final obstacles, but they presented as formidable a threat as any that Lesseps had faced. In the middle of 1863, he was scrambling to adjust. They had caught him off guard, and Lesseps and the company, so unassailable a few months earlier, were now on the defensive. In late August 1863, as summer was coming to an end, Nubar, recently made Nubar Pasha, heir to centuries of Ottoman power and Armenian pride, left for Paris to confront an arrogant Frenchman and his company.

# CHAPTER FIFTEEN

# FERDINAND FIGHTS BACK

ERDINAND DE LESSEPS had come too far to allow a pair of late arrivals to endanger his life's passion. He did not believe that either Ismail or Nubar cared one whit for the fellahin. He did not think that Abdul Aziz or his ministers cared one way or the other whether the canal was built. To his mind, the renewed flurry of opposition had one source: Great Britain. Now, however, he was in a stronger position than he had been the last time the British attacked. He was the head of a large company that was enmeshed in French society. The canal had become part of public life, and its fate was tethered to French honor.

At least that was what Lesseps intended to argue. Rather than debate the morality of the corvée (an argument he knew he would lose), he tried to alter the terms of the dispute. Ismail, Nubar, and the Porte contended that the issue was forced labor and the alienation of sovereign land. Lesseps countered that the problem was the British government and the hypocritical opportunism of Ottoman ministers and their allies in Egypt and Europe. Though he did not criticize either Ismail or Abdul Aziz, he showed no restraint in attacking Nubar Pasha by name. Just as he had undermined Enfantin years earlier, Lesseps went after Nubar, and Nubar went after him.

In Paris, Lesseps addressed a meeting of the company's shareholders. The current dispute, he declared, was not a product of French intrigues. "It is English policy, which aspires to seize Egypt. In order to achieve this goal, Egypt must not be calm, content, and powerful, or able to defend herself. Egypt must be weak, troubled, impoverished, and disarmed. . . . More than ever, the creation of the Suez Canal should interest those governments who favor the neutrality of Egypt, its independence from all exclusive influences, and the maintenance of its autonomy." Egypt was the ultimate sovereign of the canal, and it would be neutral and fair to all nations. Only in England, he continued, was there any opposition, and only Lord Palmerston dismissed the universal benefits that the canal would offer. Instead of shedding false tears for the fellahin, Lesseps continued, Palmerston and Parliament should tend to injustices in their own country. As he wrote to his brother at the end of August, "The heritage of the Suez Company is not to be shared. . . . We have labored and sown, and we ourselves will reap the harvest."[1]

While Lesseps prepared his counterattack, Nubar worked with the sultan's ministers at the Porte. In late summer, the grand vizier, Fuad Pasha, issued a formal letter to Ismail. He stated that the Ottoman government favored an end to the corvée and the revocation of the concession of the lands bordering the Sweet Water Canal. While acknowledging that the company should be financially compensated for these amendments, Fuad announced that if, after six months, no agreement had been reached, work on the canal was to cease.

Granted wide authority by Ismail, supported by the Porte, and egged on by the British, Nubar relocated to Paris. Once there, he acquired another ally. In prior years, Paris had been a cold place for the enemies of the canal, but in tandem with Haussmann's continuing reinvention of the city, the political terrain changed as well. Though the emperor still had supreme power, he relaxed his hold on the political system. His health had started a slow deterioration, and after six years of uncontested autocracy, he finally agreed to allow elections for the Legislative Corps. Though he had assumed power promising Bonapartism and democracy, there had been quite a bit of the former and none of the latter. But his heart was never comfortable with severe repression; murmurs grew into rumblings, and it seemed prudent to allow for some loosening of the reins. Besides, Napoleon III did actually believe in democracy, in theory if not in practice. The elections that spring, however, were a serious blow to his pride. While the official party won a bare majority, the opposition Republicans fared extremely well. Unwilling to nullify the results, Napoleon reached out to the most prominent Republican leader, Émile Ollivier.

Challenges without were echoed by turmoil within. Napoleon's court had been dominated by the same few individuals for more than a decade. Over time, factions had developed, and though many grew rich with the booming economy, animosities hardened into cancerous hatreds. Napoleon's courtiers came to detest one another, and that had ramifications for the Suez Canal. Napoleon's cousin Prince Jérôme Napoleon was the company's official patron. But it also had a powerful enemy: Charles-Auguste, the Duc de Morny, bastard child of Hortense de Beauharnais and Talleyrand, and Louis-Napoleon's half-brother.

Morny had been pivotal to the emperor's success in 1848, and he was president of the Legislative Corps. The legislature served only as a rubber stamp until 1863, and Morny stood at the center of the graft nexus. He became one of the wealthiest men in France, after the emperor himself. He also was a leading member of the Parisian demimonde. The birth of photography had been quickly followed by the birth of modern pornography, and Morny was an early patron of this new, and risqué, business. He spent lavishly on illicit photos, and delighted in displaying them at his equally lavish parties. Like many of Napoleon's more powerful ministers, and like Napoleon himself, Morny did not distinguish between high and low culture. His desire to acquire covered a wide spectrum.

Hardly friends and barely acquainted, Morny and Ollivier became allies with Nubar in the campaign to wrest the Canal Company from Ferdinand de Lesseps. Neither man had strong convictions about the canal. Morny recognized that the canal would be built and that someone was going to get rich. He saw no reason why it should not be him. Ollivier seized on an opportunity to extend his political influence. To Nubar's credit, he was able to cultivate an alliance with both men and use them as they used him.

In the fall of 1863, both Lesseps and Nubar were in Paris, and they did everything short of fighting an actual duel. As Lesseps had done with Said's money in the 1850s, Nubar used Ismail's treasury to fund a propaganda campaign in the French press. The journalistic culture of the day did not know from objectivity, and each paper and journal was explicitly aligned with a particular faction. The assault took Lesseps and the company off guard. They had grown accustomed to kindly treatment by the French press, and the onslaught of articles questioning the probity of the company and its founders was a jarring shift. One paper, *La Semaine financière,* was especially virulent, and it singled out Lesseps personally. A typical article began, "M. de Lesseps, from blind faith, has hidden from the stockholders the necessity of obtaining the consent of the Porte," and it went on to insinuate that the company had defrauded its shareholders.

Infuriated that his honesty was being questioned, Lesseps struck back: he sued the paper for libel and rallied his supporters. The incessant negative press was having an effect. Stockholders were becoming uneasy, and the price of canal shares started to drop. That was exactly what Nubar, Morny, and Ismail had hoped. The lower the shares went, the more pressure Lesseps would face. If shareholders lost confidence in Lesseps, Morny and Ismail would have an opening. Ismail was already the single largest investor, and if Morny could persuade others to vote against Lesseps, they might be able to remove him.

As the contest intensified, it became inevitable that the emperor would get involved. Nothing of great consequence occurred in France without Napoleon's having some say, and the canal would be no exception. By the fall, the rival factions had begun to jockey for the emperor's support, and this time Lesseps was not the prohibitive favorite. Though Eugénie gave him access, that paled in comparison with the access enjoyed by Morny. Lesseps was not the only one aware of that, and he needed to placate the members of the company's Administrative Council. With characteristic bravado, he reminded them how far they had come: "We were protected by the French government during a period when we were very weak; now that we are strong we will be even better protected. We must be without fear; as for myself, I have none."[2]

At the year's end, Nubar and Lesseps were no closer to a resolution. Nubar considered Lesseps "noisy and cumbersome," and scoffed that it was impossible to negotiate with a man who acted as if "whoever was not fully for him was against him." Lesseps, in turn, was offended by the press campaign orchestrated by Nubar and outraged at what he saw as the Armenian's hypocrisy. In October, informed that no further conversations could occur until the company's board deliberated for several more weeks, Nubar chided Lesseps, "Fifteen days is of little or no importance for those of us who live peacefully in Paris, but it is a long time for the fellahin. . . . I ask you, do not delay; this is not a matter of diplomacy, but, rather, an issue of humanity." Though it was a personal letter, Lesseps made its contents known to his partisans, and he wrote an angry reply. "I have dealt with such issues every day of my life," he answered, "more than you have in yours." He also resented the attempt at manipulation. "Do not try to use such appeals when we are talking face to face," he said. "When we talk, leave aside those clever phrases that you use for the public. . . ."

By January 1864, it was evident that only the emperor could end the stalemate. The forces on both sides of the dispute were each too entrenched. Other than Napoleon, Morny was the most powerful man in France. The Canal Company, for its part, had Prince Napoleon as a patron, the empress as a guiding angel, tens of thousands of investors, and substantial public support, which, in the face of English opposition, had transformed the canal into a matter of national pride. Nubar understood the situation as well as anyone. Soon after his arrival in Paris, a man stopped him in the street. "You come to make war against de Lesseps and prevent the canal from being built," the man said. "Ah well, no! I am not rich. I have five shares, and if it were necessary, I'd put the rest of my money into this enterprise."[3]

Though no one could predict the results of a war of attrition, neither side had an interest in seeing the contest drag on indefinitely. There was no hope of compromise, largely because this was less a dispute over labor and lands than a battle for control. Recognizing that Ismail, Morny, and Nubar wanted him gone, Lesseps saw no reason to concede on the issues per se. Seeking to dispossess Lesseps and his partisans of the canal, Ismail, Morny, and Nubar were not interested in serious negotiations. Having reached an impasse, both sides turned to the emperor.

Within the space of a few weeks, each faction formally requested the emperor to arbitrate. In order to buttress his case, Morny commissioned a study from Ollivier. By allying with the popular Ollivier, Morny could argue to the emperor that the French people were beginning to doubt Lesseps and the company. Ollivier submitted a report in early January 1864, raising serious questions about the company's claims. He charged that Lesseps had consistently failed to distinguish between politics and commerce, and that he had unfortunately convinced people of influence in France that the commercial desires of the company were as important as the political issues surrounding the canal. Having carefully read the concessions, Ollivier concluded that the land grants were not essential and should revert to the sovereign control of Egypt in return for a modest indemnity; that, contrary to Lesseps's argument, the work was contingent on the approval of the Porte; and that the viceroy had the legal right to terminate the corvée.[4]

For his part, Lesseps wrote a personal appeal to the emperor to adjudicate. "Sire," he began, "The Suez Canal Company peacefully pursues its operations. . . . The results to date are due to the protection of Your Majesty, who in 1860, when the Company was beginning operations, saw to it that they were not brought to a standstill by an irregular order from the Ottoman Porte. The Company has no cause for anxiety, because being bound by agreement with the Egyptian Government, the Egyptian Government is equally bound towards the Company. . . . [But] under the pretext of a persistent refusal by the Porte to ratify the Concession, a recent letter from the Grand Vizier seeks to place the viceroy in a position to demand from the Company the surrender of some of its rights, without compensation of any kind. . . . In the same way that in 1860 Your Majesty countermanded orders which had been officially given with the object of bringing these works to nothing, the undersigned dare to hope that again on this occasion . . . the Emperor's will shall not permit the accomplishment of intentions hostile to the company; and that he deign to protect the French shareholders in the Suez Canal as much as the interests of the Egyptian government itself."[5]

But dulcet words were not enough. Lesseps knew that his adversaries were using their extensive connections to influence Napoleon. Though the emperor and his wife did not entertain quite as much as they had in the past, they still held dinners and balls, and gossip and innuendo were important weapons in the battle for the emperor's favor. Morny was a high official and a blood relation, but Lesseps still had Eugénie. Through her, he was able to arrange a private audience with the emperor at Fontainebleau, which was an exclusive compound of parks and streams trolled by an opulent imperial barge. Invited to an informal luncheon, Lesseps would have only briefly touched on the canal. He knew better than to force an issue, and the access alone, even if much of the time was spent in light, airy, and meaningless conversation, was invaluable. But he would certainly have stated, as eloquently as he could, his belief in the canal. In the end, the emperor would make his decision based not on arcane points of law or even on international politics, but on the basis of a personal affinity for the same ideals Lesseps professed.

Through all of these months, Lesseps never framed the canal as an engineering project. Instead, he described it as an expression of all that was good and just in civilization. "The Suez Canal," Lesseps said in one speech, "is not the property of a few men; it is not the property of one nation. It belongs to the human race, and it owes its birth to the aspirations of humanity, aspirations that are irresistible because they fulfill the needs of all and consecrate our epoch. . . . Steam power, railroads, electricity have shortened distances; the Occident, constrained by limits that were too constricting, suffocated by the marvels of its own industry, sought a grander outlet. It felt that it had to find it outside of wars and revolutions. . . . This is the wind of the century that blows our sails and leads us to port. . . . We have on our side right and truth. . . . How, with these with us, can we not triumph over the obstacles strewn on the road of all who strive, obstacles measured with justice by Providence in proportion to their utility and the greatness of the goal?"[6]

The emperor was a man who thought he was walking in the footsteps of greatness. He wanted his reign to mark the beginning of a golden age in human history. The fight between Lesseps and his adversaries was couched in the legalisms of concessions, land grants, and who said what to whom. Nubar and Morny and Ollivier wrote long memos justifying their positions. Lesseps did as well, but he did something they did not. He wove a romantic vision, and that was far more attractive to Napoleon than legalistic arguments pro or con. Just as Lesseps had enraptured Said to support the canal in the first place, he was able to win over the emperor. This was not sheer manipulation. Lesseps believed that the canal was an embodiment of progress. He felt that he was an instrument of Providence, and that the canal was almost inexpressibly important. Napoleon, who felt the same way about his regime, responded to Lesseps as a kindred spirit.

That did not mean that the two men became friends. Lesseps in early 1864 did not know what the emperor would decide, nor did Napoleon himself. The emperor asked his ministers to assemble data on the concessions and assess the arguments for and against the company. But, given Napoleon's character, he probably did not dwell on the details of the competing claims. He had not thought much about details when he made a failed coup attempt in 1840, or when he rose to power in a series of twists and turns between 1848 and 1851. He focused instead on fate, and on his role in fulfilling it. Asked to decide between the competing claims of the Suez Canal disputants, he would choose the one that best served France and civilization, in that order.

Meanwhile, the company's libel case against Nubar was dismissed, and the propaganda war continued. Nubar, in letters to his wife, claimed to be shocked by how shamefully he was being treated by the partisans of the canal. "My God! My God!" he wrote her. "Such calumny is a terrible thing, and men are so stupid and wicked." He in turn honed his argument for the emperor and provided Morny with ammunition.

Nubar continued to denounce the unjust use of forced labor. He calculated that, with 20,000 laborers coming, 20,000 going, and 20,000 working in the isthmus each month, there were 720,000 laborers working on the canal each year out of a total Egyptian population of less than 4 million. He claimed, correctly, that this was an extraordinary commitment. He also claimed that the resulting labor shortage meant that Egyptian agriculture was losing 36 million francs a year, and this during an international cotton boom from which Egypt could have derived even more benefit were it not for the deleterious effects of the corvée. These fellahin were paid, at most, 75 piastres. Once they had made the journey by train, by steamboat on the Nile, and then by foot across the desert, the cost to the Egyptian government was even greater. It took eight to ten days for some fellahin to get to the isthmus, and Nubar said that transportation expenses were paid for by Ismail. The fellahin themselves had to bring their own bread, or they had to purchase biscuits, salted fish, and tobacco from the company at company stores. After twenty-five days of work, they had to make the return trip. In the end, Nubar concluded, both the state and the individual fellah lost money.

Nubar's arguments were then recycled by Lord Palmerston in Parliament. "It is very much to be regretted," he stated, "in the interest both of England and of France that when both countries are much in need of cotton, 30,000 or 40,000 people who might be usefully employed in the cultivation of cotton in Egypt are occupied in digging a canal through a sandy desert and making two harbours in deep mud and shallow water. I should hope that so useless an occupation will soon be put to an end."[7]

Each move by one side brought a riposte. In early February 1864, Prince Napoleon hosted a banquet for Lesseps to celebrate the completion of the Sweet Water Canal, which had finally been extended to the Red Sea. The affair took place, appropriately enough, in the Palace of Industry, off the Champs-Élysées, and it was an unbridled demonstration of support for the company, for the board of directors, and especially for Lesseps. A panorama of the isthmus had been created for the occasion, and two dozen tables were joined to form one vast communal dining area. Lit with the latest gaslight technology, the hall was a glowing international assemblage of shareholders and friends. Senators, clergymen, magistrates, bureaucrats, diplomats, industrialists, bankers, lawyers, doctors, and artists, as well as members of the press corps, gathered to listen first to music played by the gendarmes of the Imperial Guard and then to toast after toast. It was the type of evening that the critics of bourgeois culture detested. Men of society stood and portentously hailed Lesseps and the great venture of Suez. During dessert, the head of the Paris Chamber of Commerce rose. "Gentlemen! In the name of the Banquet Commission, I have the honor of presenting a toast. To His Imperial Majesty, the Emperor! To the Protector of noble and grand enterprises! . . . To the piercing of the Isthmus of Suez. Gentlemen, let us thank the Emperor. . . . To the health of the Emperor, and to the health of Her Imperial Majesty the Empress, and to the Prince Imperial!"

Then Plon-Plon himself lumbered to the podium. "To the company!" he began, and proceeded to talk in his crude, rambling way for nearly two hours. "Our enemies want to portray our success as a moral failure, and it is to respond to these attacks that we are gathered here this evening." He had visited Egypt and witnessed a French company creating a monument in the desert to rival those of the pharaohs. When the canal was completed, the whole world would be forced to agree that it benefited not just its shareholders, or France, but Egypt and all of humanity. The prince parodied the motives of Nubar and Ismail. They claimed that they weren't against progress, that they respected the "pride of France, the grand Polytechnic School." They said that they only wanted to defend the poor fellahin and the rights of the viceroy. But they were hypocrites. Nubar was the servant of an absolute master who was in turn the servant of another absolute master, the sultan. They did not care for freedom or for the rights of man, and they did not live in a culture that had ever cared for those freedoms. But, the prince continued, for all the calumny, the truth was that the company treated the fellahin better than any Egyptian or Turk had ever treated them or paid them. "The Company says to the viceroy: The corvée existed in your land long before the company showed up. The Company didn't create it. We profit from it because you have allowed it. You want us to abandon it. We're happy to, but not as an act of philanthropy. If you want us not to use the corvée, you must compensate us. And if you want to abolish slavery in your land, you have our blessing."

Finally, Plon-Plon sat down, and Lesseps rose to thank him. Not wanting to detain the restless audience, he spoke briefly. He reiterated that the canal would reunite two civilizations, and to illustrate that point, he told of how the Egyptian workers had celebrated when the Sweet Water Canal reached the Red Sea. Hand in hand with the foremen, who were mostly French or European, the workers chanted, "The Christians are also children of God. They are our brothers." Lesseps spoke once again of progress and humanity. He made a toast to the memory of Said; to the current viceroy, Ismail; and to everyone present in the hall that evening. True or not, the story of Muslims saluting their Christian foremen played to the hope and desire of the audience that they were engaged in something greater than an engineering project. Lesseps told them that the canal lay at the end of a centuries-old continuum, stretching from the pharaohs and then the New Testament, and through history to the present. Not only was it the fulfillment of ancient ambition, it was also the junction where Christian and Muslim would end their long conflict and join hands for the betterment of mankind.[8]

Nubar and Morny had trouble countering this rhetoric. In theory, they could have used the corvée as the centerpiece of a florid attack on the company, but that would have backfired. The Egyptian government depended on the system, and it would have been snared by the same arguments. Rather than focus on the morality of forced labor, Nubar and Morny restricted themselves to challenging the company's legal rights.

In March, the contending parties officially submitted their grievances to an arbitration commission appointed by the emperor. The commission would assess the 1856 statement issued by Said about the corvée, and address the issue of an indemnity for the company. It also would examine the land grants. It planned to conduct hearings, and to make a final recommendation to the emperor by the end of June.

As the commission started its work, Lesseps, recognizing that the emperor now had ultimate power to determine the fate of the company, increased the frequency of his letters to Eugénie. In public, he announced that the company was eager to end the corvée, but that, in light of Said's promise to furnish the canal with laborers, it had the right to be compensated. Nubar contended that in the absence of the Porte's approval, none of the prior agreements had the status of contracts. Therefore, the company had no right to compensation if the current viceroy, Ismail, changed his mind and made a decision that honored universal notions of morality.[9]

But the opponents of the canal were at a disadvantage. They were allies only because they shared an enemy. Morny's motive was greed; Ollivier wanted to gain influence in France; Ismail hoped to gain control of the company; and Nubar sought to enhance the power of Egypt. They did not have a coherent ideology. Meanwhile, Lesseps and the company had a philosophy that simultaneously reflected the ideals of nineteenth-century society and yet was compatible with the drive for profit, the quest for fame, the love of country, and the hunger for technological innovation. Until the emperor rendered his decision, both sides wondered anxiously about the outcome. In truth, it was never a fair fight.

On July 6, 1864, Napoleon issued his ruling. It was an absolute victory for Lesseps and the company. The commission had made its final recommendations several weeks earlier, and though it was purportedly a neutral body, it did not deliver a neutral assessment. Nor did the emperor. He accepted the company's argument that the concession of 1856 had the status of a binding contract, and he agreed with the company's contention that Said's promise to supply workers also constituted a contractual commitment. Even though the viceroy had the authority to alter the terms of the concessions granted by his predecessor, the company had the right to an indemnity.

According to the imperial sentence, the Egyptian government would have to pay the company thirty-eight million francs in order to compensate it for ending the corvée. Though the sum could be paid out over several years, it was still an enormous amount of money. The other half of the award covered the lands on either side of the Sweet Water Canal, which had originally been promised to the company. The idea in 1856 had been that the newly irrigated lands would generate an income that would offset the expense of digging the Sweet Water Canal. Napoleon ruled that if Ismail wished to revoke that promise he would have to pay the company thirty million francs, in annual installments. The government was required to give the company a bulk payment of ten million francs, and another six million for navigation rights. The total amount: eighty-four million francs.

The contestants were informed of the decision, but it was more than a month before the full text was released. During that time, there was some uncertainty about what the award actually meant. The payout schedule was complicated, and the language of the decision was legalese. For several weeks, shares of the company traded lower on the Paris Bourse, because at first blush it seemed as if the company would lose its land rights. But once it became clear that in return for those rights there would be a substantial monetary payment, the shares recovered. By the end of summer, there was no longer any doubt about the company's victory. Nubar was bitter. He believed that the elites of Europe had once again found a way to profit at Egypt's expense, just as they had throughout the Near East and Asia. For his futile efforts, Nubar was awarded the Legion of Honor by Napoleon, on Morny's urging. "It is a heavy price to pay for a medal and it weighs me down," he told Morny. He was right to be dispirited. Eighty-four million francs equaled the entire annual budget of the Egyptian government.[10]

Yet Ismail did not seem quite so troubled. Having placed himself in the hands of Napoleon, he agreed to abide by the decision, even though the English government urged him to contest the settlement and argued that the emperor of France had overstepped his authority by intervening in the sovereign affairs of the Egyptian viceroy and the Ottoman sultan. Ismail believed that the canal, more than ever, belonged to Egypt. He was the largest shareholder. As a result of the arbitration award, he now had ownership of the Sweet Water Canal and its lands, and he was entitled by earlier agreements to 15 percent of the profits once the actual canal was opened to ship traffic.

In addition, Egypt was still enjoying the benefits of high cotton prices, and Ismail felt that he would be able to meet the financial demands of the settlement. Cotton wealth had led to inflation in Egypt, and Ismail was confident that would enable his treasury to fund both the indemnity and the public works he had authorized. Optimistic about his future, he took a public loan, negotiated with the Oppenheim brothers, for nearly one hundred million francs (4.8 million British pounds plus a healthy 15 percent commission for the Oppenheims). With interest, this would ultimately cost Egypt twice as much. It seemed a reasonable, though onerous, commitment at the time. But it was the first step toward European control of Egypt's finances, and the first of many loans that Ismail incurred. Bit by bit,

Egypt slipped out of his hands and into the pockets of European bankers and businessmen.[11]

In public, Lesseps did not exult. Ismail, Nubar, and Morny had been dealt a conclusive defeat, but there was no reason to humiliate them. The company still needed the cooperation of Ismail, and Lesseps respected hierarchy. Ismail was the viceroy, and as such deserved deference. In the statements Lesseps issued announcing his satisfaction with the emperor's decision, he stressed that there was much work left to be done and that it was now time to turn away from disputes over contracts and toward the formidable task of completing the canal.

Just before Lesseps returned to Egypt that fall, Maxime Du Camp, Flaubert's traveling companion years before in Egypt, took a short trip outside of Paris to visit an old, frail man who was on the verge of death. Prosper Enfantin had lived a peculiar life. He had become richer than most, and he had left his mark even on those who had rejected him. Now, looking back, he had come to accept his limitations. "In my hands," he told Du Camp the week before he died, "the canal affair was a failure. I did not have the necessary flexibility to deal with all of the adversities, to fight simultaneously in Cairo, London, and Constantinople. . . . In order to succeed, one must have, like Lesseps, a devil's determination and ardor that doesn't know fatigue or obstacles. By the grace of God, it is Lesseps who will marry the two seas. . . . It matters little that the old Prosper Enfantin was defeated by deception; it matters little that his children had their hopes betrayed. But it matters greatly that the Suez Canal will be pierced, and it will be. And for that reason, I thank Lesseps, and I bless him."[12]

It had been a good summer for Lesseps. With Napoleon's ruling, he defeated his last and most serious challenge. But with Enfantin's blessing, he achieved his most improbable victory. Now all that remained between his vision and reality was sixty million cubic meters of sand.

# CHAPTER SIXTEEN

# MEN AND MACHINES

I T WOULD TAKE another year and a half for the company, the viceroy, and the sultan to arrive at a final agreement. The emperor's decision had only established the overall framework. Though the ending of the corvée was straightforward, many details were unresolved. Ismail and the company promised to honor the imperial arbitration, but they were far apart on exactly which acres would be returned to the Egyptian government, when they would be transferred, what lands the company could draw revenue from and until what date, and how long the company could enjoy tax-free use of the Sweet Water Canal. To complicate matters, England remained hostile. Even the most truculent British official realized that the completion of the canal was a foregone conclusion, but having opposed its creation for so long, they were unwilling to concede defeat. Unable to halt progress, they could still make the process difficult.

The diplomatic focus shifted to Constantinople. Ismail had all along intended to alter his legal status in the Ottoman Empire. He wanted the right to pass his mantle to his eldest son, and he wanted more autonomy. The imperial arbitration provided the opportunity to win these concessions from the sultan. Lesseps and the company directors felt that it was symbolically important to obtain the sultan's consent in writing. The canal was, after all, in the Ottoman Empire, and it would change the commercial and strategic balance of the eastern Mediterranean. Though the silence of the sultan could not prevent the canal from being opened, it could hurt the company in the future. What if a new sultan came to power and tried to assert control over the canal? Large companies do not like political uncertainty, or even the prospect of it. It was better to get the sultan's approval sooner rather than taking the risk of not being able to secure it later.

Once again, Ismail relied on Nubar as his emissary. Frustrated by the outcome in Paris, Nubar welcomed the challenge of Constantinople. He knew the terrain, and he enjoyed the delicate indirectness of negotiations. He also understood the culture of bribes, and he was given wide latitude by Ismail to spread the wealth. The company was less adept at this game, but it was in a much stronger position than it had been in years past. Lesseps and Ismail were more or less on the same side. Though they still disagreed about certain details, they both wanted to see the 1864 arbitration ratified by the sultan. The company also enjoyed the explicit support of the French government. Lesseps and his representatives no longer needed to insinuate that the emperor favored the canal: he did, and everyone knew it. The goal for both the company and Ismail was to get the sultan to agree to a change in the viceroy's status and then sign off on the arbitration. The British, interested only in obstruction, could not match the combined efforts of the company, the French government, and Nubar.[1]

In the fall of 1864, the final outcome of these maneuvers was unclear. At various moments, the situation seemed to be reaching a new crisis, but all the while, work on the canal accelerated. The outstanding issues were not settled until February 1866, but during that interval of eighteen months, Lesseps expended less energy on diplomacy than at any point in the previous decade. Instead, he focused on building the canal as quickly as possible.

That was imperative, because for all of the talk, the physical work had barely begun. At the end of 1863, Hardon was finally eased out of his position as general contractor. He had spent tens of millions of francs with not much to show for it. He also presented the company with a large bill, in excess of half a million francs, and he was fortunate that Lesseps preferred to pay him to go away rather than try to penalize him for the many false starts. By the time the general assembly of shareholders was asked to endorse the settlement reached between Hardon and the company, new contractors had been hired, and their success removed the sting of Hardon's failure. At best, the mistakes made on his watch led to the innovations that allowed the canal to be built; at worst, he had wasted time and money and jeopardized the entire endeavor.

As Hardon receded, Voisin Bey took over. He inherited a haphazard organization of isolated stations in the isthmus and transformed it into a modern company, with a clearly defined chain of command. He also helped find two contractors who made the subsequent completion of the canal possible. By early 1864, the company was beginning to lessen its dependency on human labor. In part, that was a logical response to Ismail's stated intention to end the corvée. Even with an indemnity, the cost of tens of thousands of noncorvée workers would be prohibitive. The company would have to provide salaries and incentives to get laborers to the isthmus, and it might even need to pay for some of the transport. But even without the tumult surrounding the corvée, the company realized that the canal could not be completed at anywhere near its original budget if it relied on thousands of fellahin digging by hand. This was not because the fellahin were incapable. It was simple arithmetic. Unless the number of workers was doubled or tripled, they could only remove so much sand so quickly.

The solution was to mechanize. By the middle of the nineteenth century, engineering projects throughout Europe, the United States, and Latin America were increasingly relying on a combination of men and machines. Steam- and coal-powered contraptions helped carve out tunnels, deepen mines, dredge harbors, and remove debris. There were entire businesses and factories devoted to the production of machines, but most of these new tools were designed and developed piece by piece and tailored to the task at hand. In France, it was common to produce heavy, mechanized equipment to specification; mass production of machine tools was several years off. As a result, when the Canal Company decided to make the transition to mechanical dredgers and elevators, it had to find someone who could both oversee their installation and use, and actually design and construct them in the first place.

In December 1863, Paul Borel and Alexandre Lavalley formed a partnership, called Borel, Lavalley, and Company. Their corporate title may have lacked imagination, but their actual work did not. Like Voisin and so many others, they were graduates of the Polytechnic School, where they absorbed the ideals of progress, industry, and civilization that had infused the project from the start. They both had extensive experience working on railway lines. Borel, an intense-looking man with a penetrating gaze, was born in Marseille, and had learned his trade working on the Bordeaux-to-Tours line, and then on the important link between Lyon and Marseille. Fascinated by machinery, he formed a joint venture with several partners to manufacture train engines. The younger, genial-looking Lavalley, called by Lesseps "the most skillful engineer in Europe," spent many years in England as an apprentice and then a designer of specialized machines. He became an expert at metallurgy, understood the workings of forges, studied the latest steam-power technology, and then applied his knowledge to customizing locomotives. His dexterity had taken him far afield. He designed lighthouses on the Black Sea and the Baltic; he created machines to bore a tunnel in Lithuania; and he fashioned dredgers to deepen ports in Russia.[2]

These were precisely the skills that the Suez Canal Company needed to move to the next phase of construction. Though there were a handful of dredgers in place at Port Said, most of these had been designed for work on the Nile. For work farther south, in the isthmus, different machines would be needed. At Lake Manzala and Lake Tim-sah, some of the excavation would be done in shallow water. The plateaus of El-Guisr, the Serapeum, and Chalufa were of different heights, and the mix of rock, soil, gypsum, and sand varied from place to place. Each excavation elevator had to be tailored specifically to account for these differences. Train tracks had to be configured to accommodate the machines. Some dredgers would be anchored on barges and used to remove the mud and sand from the bottom of the canal, whereas others would be placed on tracks along the banks of the canal. The angles for the conveyor belts and the depth of the elevators would differ, depending on where they were used. The amount of weight each bucket could carry depended on the consistency of the earth, which varied greatly throughout the isthmus.

The result was that more than a dozen different types of machines had to be designed, constructed, and then installed. Many of these came in four, five, or six different sizes. The initial prototype often went through several models. Once they were actually deployed, the new machines were subject to tinkering or redesigns. Some of that could be done in the workshops at the isthmus stations, but some of the cus-tomization had to be completed in France. For instance, one dredger meant to remove the fine sand near the Mediterranean coast kept breaking when the sand and mud clogged the works. The only solution was to rinse the machine regularly with water, and that meant redesign-ing the dredger and constructing and installing a water pump. There were countless adjustments like this, each of which slowed progress.

For five years starting in 1864, Borel and Lavalley furnished the company with machines. They devised them, tested them, delivered them, installed them, and, in conjunction with Voisin, hired people to use them, fix them, and maintain them. Ultimately, there would be nearly three hundred such machines, most of them powered by coal, some by steam. Borel oversaw operations from his headquarters in Paris, while Lavalley spent most of his time in the canal zone. They obtained contracts from the company for distinct phases of the work. Rates were determined by setting a price for the excavation of one cubic meter, and the price varied depending on the consistency of the soil. For instance, the excavation of El-Guisr between Lake Balah and Lake Timsah, a distance of nine miles, contained approximately nine million cubic meters, at a cost of 2.5 francs per cubic meter, as opposed to a cost of only 1.99 francs per cubic meter for the relatively loose and easy-to-remove soil found at the southern end of the canal.

Negotiations between contractors and companies are often acrimonious, but Borel, Lavalley, Voisin, and Lesseps appear to have had a collegial, respectful rapport. Of the seventy-four million cubic meters removed from the main canal, Borel and Lavalley were responsible for more than 75 percent of the total. Most of the work was performed between 1867 and 1869. It took years for the new machines to be put to most efficient use, but by 1864, Lesseps and the company directors recognized that Borel and Lavalley would be able to deliver what others had not.

But Borel and Lavalley did not have the expertise to solve the problem of the Port Said jetty. That was left to the Dussaud brothers. The four siblings from Marseille had used innovative techniques for the construction of jetties in Algiers and Cherbourg. Experts at masonry, they had experimented with new ways of mixing concrete, and they adapted these in order to solve the riddle of Port Said. Beginning in late 1864, the Dussaud brothers developed a system for creating the immense blocks needed for the two long jetties into the sea. One jetty was to be nearly two miles long, the other a mile and a half, and between them, they would enclose a triangular harbor area of 550 acres where ships entering and leaving the canal could safely anchor. Each of the blocks for the jetties weighed more than twenty tons, which were produced on an assembly line on an island near the town. Mechanical elevators poured sand into rudimentary cement mills, and then lime imported from Theil, in southern France, was added along with salt water. The lime-and-sand cement mixture was poured into large wooden frames and spread evenly to form blocks. The blocks were then left to bake in the sun for two months; once hardened, they were lifted by hydraulic cranes onto trucks and rolled onto custom-built barges that were equipped with wooden runners that held the stone slabs at an angle. The barges transported the blocks out to sea, and then the stops on the runners were released to allow the slabs to slide into the water, one after another. In the end, the jetties contained thirty thousand blocks.[3]

There were other engineers, of course, as well as other contractors and subcontractors. An Englishman named William Aiton oversaw a portion of the excavation around Port Said, and a French entrepreneur named Alphonse Courveux helped tackle the challenge of cutting through the plateau at El-Guisr. But by 1866, Borel and Lavalley had bought out Aiton, and they served as the de-facto general contractors for the last half of the work, answerable only to Voisin Bey and Lesseps. They drew to the canal many of the most promising young engineers in Europe, who came for the adventure and for the experience. Voisin also assembled a skilled team of regional chief engineers, men such as Félix Laroche at Port Said and Eugène Larousse at the port of Suez, also graduates of the Polytechnic. Their belief in science extended beyond machines. They also applied the philosophies of the Polytechnic to human organization. They thought that, just as machines were composed of multiple moving parts working in harmony, the work as a whole should consist of units and subunits working jointly for a common goal established beforehand and implemented rigorously. As a result, the Suez Canal Company embodied the notion of scientific management long before that term was coined.

With the arrival of Borel, Lavalley, and the Dussauds, working for the Suez Canal Company became one of the most desirable jobs in Europe. Ambitious young engineers knew that having the canal on their résumés would be worth a fortune in future prestige and subsequent jobs. The end of the corvée increased the need for semiskilled European workers and foremen, and the mechanization opened up countless opportunities for apprentices. In 1863, there had been tens of thousands of fellahin and only a few thousand Europeans. By the mid-1860s, there were ten thousand Arab and Egyptian residents in the various towns and settlements along the canal, and more than eight thousand Europeans. The non-Europeans were a mix of fellahin,

Bedouins, people from Cairo or Alexandria, and Syrians. Several thousand were employed directly by the company, while others ran stores or coffeehouses and provided daily necessities. As time went on, workers also brought their families, and the resident population included wives and children, of both Europeans and non-Europeans alike.[4]

In an age when someone like Lesseps could make himself into one of the most prominent citizens of the day, there were thousands of other ambitious men who believed the same thing about themselves. Many of them, finding Europe too competitive, traveled far, to the new colonies of France in North Africa or of England in India, South Africa, and the Far East. One of the best ways to advance in society, then as now, was to become involved in important projects run by important people, and the Suez Canal by the mid-1860s was the largest, most expensive, and most scrutinized engineering project in the world. A mason or a machinist in his twenties, a hydraulic engineer or a metallurgist in his early thirties might aspire to oversee the construction of a bridge or railroad or tunnel, but the only way to get there was to work steadily, make contact with other prominent engineers, and establish a reputation. The canal provided that, and not just for engineers. It offered an opportunity for accountants, shopkeepers, clerks, doctors, carpenters, surveyors, topographers, barge captains, blacksmiths, masons, stonecutters, train conductors, cooks, and telegraph operators, as well as for hundreds of minor functionaries who went to work for the company. Even so, joining the Canal Company was a major commitment for someone living in rural France or England or Italy. The grandeur of the endeavor, the romance of the desert, and the good publicity helped the company. Yet people were drawn not just by the work itself and by the prospect of being involved in something monumental, but by the salaries, which, from the director general of the works on down, were considerably higher than anywhere in Europe at the time.

What was true for skilled workers was also true for the laborers who replaced the fellahin of the corvée. Given the cotton boom and inflation in Egypt, salaries for unskilled workers had to be high enough to draw them to the isthmus. In 1865, Ismail provided the company with several thousand "soldiers," who were mostly ill-trained and were in essence corvée labor by another name. But as the work became more mechanized, there was less need for them. Soon, the company was hiring workers with minimal assistance from the Egyptian government.

Ismail had wanted to end the corvée in order to increase his power.

Instead, his influence over the canal zone decreased. As long as the corvée was in use, government officials were stationed throughout the isthmus to keep an eye on the fellahin in conjunction with company officials. But once it ceased, there was no need for Ismail's agents. The concessions had given the company wide latitude to govern the area during the construction, and though each district had an Egyptian administrator, after the corvée the company began to act as a quasi-autonomous state. There was not much Ismail could do about that. Before the canal, there had been no governmental presence in the isthmus outside of the port of Suez and the western side of Lake Manzala. Ismail could and did designate new administrative zones around the major canal settlements, but the towns were being built by the company; the company owned the stores; the company paid for and owned the rail links; and until a final settlement was reached, the company controlled the Sweet Water Canal and all transport from Cairo. In 1865, Ismail authorized and funded the construction of a railroad from Cairo to Ismailia, but it would be years before that was complete. In the interim, as the population of the canal towns and settlements rose above twenty thousand, the company took on the functions of a government. It policed the towns, and it kept the peace. And slowly, its administrators began to assert their prerogatives at the expense of Egyptian officials.

In April 1864, the governor of Damietta lodged a complaint with Laroche, the chief engineer of Port Said. The district of Damietta included the lands around Lake Manzala, as well as the lake itself. It had come to the governor's attention that fishermen had begun to fish in the newly excavated canal in the eastern part of the lake. That disturbed the governor. He did not object to the fishing per se; after all, the few people who lived in the area had been trawling the lake with their nets since antiquity. But he complained that the government was owed duties on the catch, and he demanded, politely, that either the fishermen employed by the company or the company itself pay the taxes levied on all fishing in Egypt.

Though Laroche was the company agent in charge of the region, he recognized that this type of dispute was, as it were, above his pay grade. He referred the complaint to the director general, Voisin Bey, and Voisin Bey immediately referred the matter to Lesseps. Lesseps wasted no time in sending a somewhat impolite response, and forwarded copies not only to the provincial governor who had raised the matter in

the first place but to the French consul general in Alexandria. Anticipating that this was just the sort of thing that sovereign governments
can become agitated about, he wanted his reasons understood.

Though he did not say so in his letter, Lesseps had a sentimental
attachment to the fishermen of the lake. They were, he remarked in a
speech a short while later, "a race of men who do not at all resemble
the rest of the inhabitants of Egypt. It is believed that they are
descended from ancient shepherds. M. Mariette, the wise explorer of
Egyptian antiquities, has unearthed in this area statues that date from
the time of Joseph, and these had similar physiognomy and special
characteristics that can be observed in the current residents of the
lake." These were the same people who had developed the only effective method of excavating the mud of the lake, and whose labor on the
canal had been vital until the dredgers designed by Lavalley supplanted them. But whether or not they were directly descended from
Biblical shepherds, and whether or not they had aided the canal's construction and provided the residents of Port Said with food, Lesseps
did not think that their labor belonged to the Egyptian state. Instead,
he rejected the governor's demand. The fishermen, he said, were working for the company, or, at the least, working as independent contractors supplying the company with fish. According to his reading of the
concession of 1856, the company was entitled to "the use of all land
and waterways not belonging to private individuals." As a result, no
taxes were owed the Egyptian government, and he would not allow the
governor of Damietta to collect them.[5]

Lesseps may have been within his rights to reject claims that the
company pay duties, but it was a considerable stretch to assert that
Egyptians who worked for the company did not owe taxes. Though the
local fishermen must have celebrated the company's decision, Lesseps
was blatantly infringing on the sovereignty of the Egyptian state. Well
before the canal was finished, the fears of Muhammad Ali were coming true. A European company, backed by a powerful European government, was starting to erode the autonomy of Egypt. Other states
around the globe had already discovered that European commercial
ventures could be lethal to local autonomy. Said and then Ismail
looked to the canal as a conduit to future Egyptian greatness. The
closer the canal came to completion, the more unlikely that seemed.

Fishing rights were a minor matter, but the willingness of Lesseps
to dismiss Egyptian sovereignty was not. The company was becoming

more autonomous and more imperious. Though Lesseps had always challenged the right of the viceroy to alter the concessions without compensation, he had treated both Said and Ismail as authorities to be respected. In 1854, he had simply been a well-connected private citizen with a scheme. A decade later, he was the president of a multinational company that employed thousands of workers and was at the center of international politics and finance. He was regularly lauded at share-holder meetings and banquets, and toasted as one of the true geniuses of the day. At a banquet in 1865, he was hailed as "the collaborator of God" whose name would live forever. He had never suffered from excess humility, and it would have taken a preternaturally modest soul not to be puffed up by all this praise. The more power he acquired, the less he was willing to defer to Ismail. The more power the company acquired, the less it was willing to defer to the Egyptian government. The fishing rights in and of themselves were trivial, but what began with perch in Lake Manzala ended not many years later with a viceroy exiled, a treasury bankrupt, a company wealthy beyond all measure, and an Egypt governed and occupied by a foreign power.

Throughout this period, Lesseps maintained his peripatetic life-style. He rarely stayed in one place for more than a few months, though he had homes in central Paris, near the company's offices, and a comfortable residence in Ismailia. Whenever he was in the canal zone, he spent days inspecting the work, and he enjoyed poring over plans with his engineers. When he was in France, he insisted on receiv-ing detailed reports, prepared specifically for him. Nothing was too insignificant to attract his attention, and the files of the company bulge with instructions from him to district chiefs, foremen, and local super-visors. He wanted to know precisely what was going on where. He also was an active administrator of the company's business affairs, and he kept careful tabs on the finances. As much as he may have wanted to control all aspects of the project, however, he was too busy to decide everything, and he had good instincts about when to defer to the wis-dom of his executives and engineers and when to demand a different course of action. And at no point did he have the luxury to focus pri-marily on the logistics of the construction. He was the public face of the Canal Company, and he was the only one who could effectively promote the enterprise.

By 1865, the canal began to attract more visitors. These ranged from curiosity seekers making a detour from the Nile antiquities and the

Holy Land, to journalists from Europe and the Americas looking for a good story, to business leaders trying to gauge when the canal would be open for trade. Commercial delegations were greeted with as much pomp and circumstance as the company could provide. Leaving a good impression was vital to future business, and the company developed a model tour for visiting dignitaries. In April 1865, the company hosted a delegation from the New York City Chamber of Commerce, along with representatives from Austria, Greece, Belgium, Russia, Persia, and the Low Countries. With the Civil War ending, American merchants looked once again to opportunities abroad, and the arrival of the Americans drew the attention of company officials. The British still accounted for most of world trade, but the United States was rightly perceived as an emerging power. Hosted by Lesseps himself, the delegates were transported from Cairo on trains outfitted with the latest luxuries. They were given tours by both Borel and Lavalley, and they were treated to a feast in Ismailia. The head of the New York Chamber, Cyrus Field (who was in the process of laying the first trans-Atlantic telegraph cable), thanked Lesseps profusely. "You have undertaken to divide two continents for the profit of all the commercial nations of the world; I fervently pray that you will soon see complete success, and that this work will rest as a monument as durable as the Pyramids because of your energy and your talent."[6]

Words such as these became increasingly commonplace, but Lesseps did not relax. In the summer of 1865, something happened that he had feared all along: there was an outbreak of cholera. From the earliest days of the company, Lesseps had insisted on maintaining an extensive, and expensive, health service. Cholera was always a risk with an inadequate water supply, and the isthmus had no consistent source of fresh water. The completion of the Sweet Water Canal remedied that, but Lesseps understood the danger of a cholera epidemic. It could strike such fear into investors and workers that progress could be crippled. Cholera was a new disease for nineteenth-century Europe. Though it had been common in India for centuries, it wasn't until 1830 that Europe suffered from its first major outbreak. It killed indiscriminately, and made a lethal pilgrimage from east to west, cutting down the likes of the German philosopher Georg Friedrich Hegel, the prime minister of France, and dockworkers in Baltimore. For decades, its cause was unknown, and physicians searched for remedies. In London, barrels of tar were burned to purify the air, and in Scotland, people were urged to drink a glass of salt water to replace the salts lost because of the pervasive diarrhea that accompanied the disease. In 1848, Paris was devastated by a cholera epidemic that killed as many as twenty thousand people. Finally, in the 1850s, physicians in London discovered that the cause was contaminated drinking water, and that led most cities to establish rigorous water standards for the first time. But fear of the disease lingered.

The epidemic that erupted in the summer of 1865 stretched from Alexandria to Port Said, and Lesseps, who had just returned to Paris, rushed back to Egypt. There was little he could actually do, but he recognized the symbolic importance of his being on the scene. The outbreak lasted for most of June and into July, and though there were quarantines, there was still loss of life. Given the lethality of the disease, the toll was mild. Several hundred Europeans were killed, including the wife of Voisin Bey. More than fifteen hundred Arabs and Egyptians also perished. The outbreak was the single largest cause of fatalities during the entire decade that it took to construct the canal.

The company's efforts at damage control yielded mixed results. Although there was no serious delay in the work, Lesseps and company officials were criticized in the French press for failing to supply timely information about the epidemic. One journal used the outbreak to critique the company, and asserted that Lesseps's opaque account of the severity of the health situation was part of a larger pattern of deception. The journal alleged that the company was substantially over budget and undercapitalized, and that the work had not proceeded nearly as far as had been claimed. Lesseps was called the lead conspirator in this plot, and in response, he sued the editors for libel and defamation.[7]

Throughout the fall, Lesseps defended his company against accusations of impropriety. It was a painful few months, aggravated by the death of his grandson, also named Ferdinand. For the second time in his life, an offspring named after him had died. His reputation was being sullied in the French press, and there still was no formal ratification of the imperial sentence of 1864. Nubar continued to be a thorn, and no matter how much Lesseps denied it in public, the company was facing a financial crisis. The machines of Borel and Lavalley were brilliantly designed, but they were also costly. The company needed the indemnity payments granted under the emperor's arbitration. Even though Ismail had terminated the corvée, until the sultan confirmed the settlement the rest of the payments were stalled. Perhaps the only good news for Lesseps was that his adversary Lord Palmerston finally succumbed to old age and died. But Lesseps was not so heartless as to celebrate the passing of such a towering presence. Palmerston had been unyielding and even unreasonable in his opposition to the canal, but he had earned the awe and respect of much of the world. The company printed an obituary honoring the prime minister on the front page of *L'Isthme de Suez*.[8]

Momentum shifted once again, however, when Nubar finally secured an agreement with the Porte in 1866. Under its terms, Ismail was given a new title. The question of what title had been a major stumbling block. The sultan wanted one that did not suggest an increase of Egyptian autonomy at the expense of the Ottoman Empire. The viceroy wanted one that did. They compromised by making one up. Under the terms of the sultan's decree, Ismail would become Khedive Ismail. It was a unique title; no one else in the world had it. The word was a neologism adopted from an archaic Persian honorific meaning "lord" or "master." As one commentator later said, "For Ismail, it meant everything; for the Sultan, nothing."[9] Ismail had spent a mint in bribes, and in return he had gained only the coveted right to pass his throne to his son while acquiring a title that did not alter his actual status. He had inherited from his father and grandfather a deep desire for direct inheritance, and for that, he had expended energy and money. At the time, he was satisfied with the result, but in retrospect he came to regret that he had fought the wrong fight.

The sultan's firman in March 1866 confirmed the rights of the company under the 1864 award, and ratified a new convention signed by the company and the khedive. Though it did not depart from the emperor's arbitration, it was more specific, and it laid out the prerogatives of the Egyptian government in the canal zone. These included setting up a postal-and-telegraph service and appointing administrators for the non-Europeans of the canal cities and settlements. The convention also stipulated that the government would collect customs duties for all non-canal land traffic on the isthmus, and it established procedures for the arbitration of disputes between the Egyptian government and the company.

This issue of legal jurisdiction was not fully resolved by the 1866 convention, however, and for the next decade, Nubar and Ismail fought to reform the entire judicial system of capitulations. In the meantime, the company continued its gradual acquisition of sovereign powers, leaving the Egyptian government to police the non-Europeans. The company governed the Europeans of Port Said, Ismailia, and Suez, while Egypt administered the rest. Two societies evolved on the isthmus. One of these was affluent and European; the other was working-class and Egyptian. One consisted of elegant homes on wide avenues laid out on a grid pattern, the other of mud huts and smaller buildings haphazardly connected by alleys and culs-de-sac. The Europeans of Ismailia lived in grand mansions made from imported materials. The home that Lesseps built for himself near the shores of Lake Timsah looked much like a mid-nineteenth-century French country house, surrounded by trees and cooled during the day by shade and breezes. The Arabs who worked for the company created a ramshackle community that looked much like a Nile Delta town. The two cultures only intermittently overlapped. The Europeans went to church on Sunday mornings, but lived amid the daily calls to prayer. The two communities spoke different languages, and knew different histories, yet both drew their livelihood from the canal and the company. The two worlds coexisted, separate but symbiotic, until the middle of the twentieth century.[10]

Though Lesseps won his libel suit against the journals that had lambasted him in the summer and fall of 1865, they had drawn attention to a disturbing fact. The company would not be able to complete the canal at its current level of financing. The best-case scenario was that the work would be finished in the middle of 1868, but only if none of the machines malfunctioned and nothing unexpected happened once the excavation reached the southern ridges of Chalufa. That was unlikely. Large projects rarely proceed without delays and glitches, and nothing in the previous years suggested that the Suez Canal would be different. The company had ways to disguise the amount of money it had relative to the amount it was spending, but it could not do that forever. Each annual meeting of the shareholders required an account of what was being spent, on each dredger, on each contract to excavate, and on each block for the jetties at Port Said. At some point in 1867, the company was going to need an infusion of cash. It could issue more shares or bonds, but one way or another, additional money would have to be found. If not, the company would be left with a half-dug ditch, thousands of angry shareholders, and a fiasco almost as impressive as the canal itself.

# CHAPTER SEVENTEEN

# THE CANAL GOES TO PARIS

COMPANIES THAT CONTINUALLY promise that their forecasts will be met tend to be punished by investors and the press when that doesn't happen. For years, Lesseps had indulged in Panglossian optimism. The project was always noble; the work was always proceeding wonderfully; and success was always imminent. Inept contractors, the diehard opposition of the English prime minister, intrigues at the Porte, cost overruns, and cholera were presented as inconveniences, regrettable but not particularly consequential. The onward march of history in the form of the canal was all but inevitable, and though the details were sometimes bothersome, the outcome was certain. This attitude permeated the company and served it well, for a time. Obstacles were brushed off, and the relentless focus on the goal soothed anxious investors. Then the company ran out of money.

Even with the indemnities to be paid by Ismail, the company needed more capital. The question was how to obtain it. The shares were already trading below the offering price of five hundred francs, and issuing more would have further diluted their value. The company's stockholders had been loyal, but they were also nervous about their investment. An announcement that more capital was needed to finish the project might undermine confidence and send the shares into a tailspin. Rather than penalizing the initial shareholders by allowing newcomers to buy a piece of the Canal Company for a fraction of the original cost, the company elected to raise the money through a public loan.

The goal was to generate an additional hundred million francs in the form of 333,333 bonds. Each three-hundred-franc *obligation* would pay interest at a variable rate not to exceed twenty-five francs per year, and the bonds themselves could eventually be redeemed at five hundred francs apiece, fifty years from the date of issue. No matter how adroitly the loan was presented, the company was concerned about the potential for bad publicity. More than ever, the Suez Canal Company was linked to national prestige. If it faltered, it would be criticized not just for organizational failures, but for staining the image of France.[1]

Lesseps recognized that the canal was no longer his own personal grail. It had become a *cause célèbre* throughout Europe. That is what he had wanted all along, but with the change came a new set of liabilities. Gone were the days when he could just write letters to the directors and patrons of the company and quietly resolve whatever issues there were. Every move was now scrutinized by a domestic and international press that treated the canal as an ongoing story of interest to readers. So many visitors were traveling to the canal zone that any problem would quickly be discovered and word of it would spread. Raising a loan qualified as a major story. Hoping to dilute the impact, Lesseps timed the announcement to coincide with another major story.

In the spring and summer of 1867, Paris was home to the Universal Exposition of Art and Industry. It was the most expensive, elaborate festival of its sort that the world had ever seen. The first of these expositions had been held at London's Crystal Palace in 1851. That was followed by the Paris Exposition in 1855, and twelve years later, Napoleon III decided to outdo both the English and himself. The Universal Exposition was an early version of what would later become the world's fair, and it combined elements of a medieval festival, a Greek bacchanalia, and the nineteenth-century cult of progress. The 1867 exposition was several years in the making, and Napoleon wanted the festival to establish France as the leading nation of Europe. He saw England as his rival for that position. England was the leading industrial power—innovative, productive, and energetic. But if Napoleon had paid more attention to his eastern frontier, he would have noticed that Germany was becoming more powerful by the month. The 1867 exposition was designed to show that France rivaled England, but given subsequent events, it wasn't England that France needed to worry about.

The setting was the Champ-de-Mars, on the Left Bank of the Seine, near where the Eiffel Tower would rise over a much-changed Paris during the exposition of 1889. Covering several acres, multiple pavilions, and thousands of exhibits, the event was a masterpiece of organization. The arts were prominently represented. The Parisian establishment did not know it, but its days were as numbered as those of the Second Empire. Already, Édouard Manet had bolted from the Salon of 1863 when his *Déjeuner sur l'herbe,* with its naked woman seated amid clothed men on a picnic, was rejected. After that, the emperor permitted the formation of an alternate Salon des Refusés. A disparate group of painters, then dismissed as Realists, began interpreting scenes of daily life in a looser, more impressionistic style. At the academies, meanwhile, Ingres and Jean-Léon Gérôme remained powerful, and they trained their students to scorn the plebeian subjects of the Realists and to produce meticulous tableaus inspired by the Bible, by the Athens of Pericles, by Ovid or *The Arabian Nights* or the myths of antiquity. Though art sometimes reflects life, the art world rarely does. In this case, in this period, it did: the disintegration of the French art establishment neatly mirrored the fraying of the Second Empire.

But that summer of 1867, the end of the empire seemed distant. Paris appeared to be the capital of the future, and the emperor and the empress presided over the exposition confident that many years of success lay ahead. The theme of the Universal Exposition was simple: science and industry could make anything possible. One observer called the fair "the last success of the Saint-Simonians," and though it may not have been the last, it was certainly the most extravagant. All of the luminaries of France participated in some fashion. The writer and teacher Augustin Sainte-Beuve, who had famously dubbed Napoleon III "Saint-Simon on horseback," lent his pen, as did the arch-critic of the empire, Victor Hugo. Alexandre Dumas, Ernest Renan, and Théophile Gautier all wrote paeans, guides, poems, or papers promoting the exposition and France. Each of them had been influenced by images of the Near East or had gone themselves to Egypt, Algeria, or the Holy Land and used what they saw and smelled and heard as fodder for future writings and future works. The exposition worshiped the technology of the West, but its promoters had ingested the culture of the Near East.

Hugo was still in exile on the isle of Guernsey in the English Channel, and he still loathed Napoleon. But love of country trumped disgust at its ruler, and he wrote the introduction to the official guide published by the organizers. Hugo declared the event a triumph for a common European culture. He thought that technology would create a world civilization without war. "France, goodbye," he wrote. "Just as Athens became Greece, just as Rome became Christianity, you, France, become the world!" Hugo believed that France was greater than its current ruler, just as Rome had been greater than its brutal and ignorant emperors. Whatever Napoleon's limitations, the country was a light of liberty and industry leading the nations of the world to a better future. Hugo endorsed the exposition and muted his criticisms of Napoleon, while the emperor welcomed the encomium and ignored their past animosity. Enthusiastic about the festival, both men succumbed to naïveté about human nature. Contrary to the hopes and dreams of the day, technological progress meant neither an end to war nor the victory of reason and peace.

Skepticism, however, had no place at an event that lionized the human spirit. People descended on the city to enjoy the pavilions and admire what Baron Haussmann had done. The festival helped foster the modern mystique of Paris. Before the middle of the nineteenth century, few spoke of Paris as the city of light and art and refined beauty. As a result of Haussmann's renovations and the 1867 exposition, such descriptions became commonplace.

Paris was glorified for its earthly delights, but it was also criticized for them. The conspicuous consumption that Zola and Balzac skewered in their novels of the Second Empire was proudly on display at the Champ-de-Mars that summer. Wilfrid Scawen Blunt, the English writer and poet who spent many years in Egypt and married Lord Byron's granddaughter, rhapsodized about the city that the exposition unveiled. "Paris! What magic lived for us in those two syllables! What a picture they evoked of vanity and profane delights, of triumph in the world and the romance of pleasure! How great, how terrible a name was hers, the fair imperial harlot of civilized humanity!"[2]

Eleven million visitors paid a total of twenty-seven million francs to survey the marvels of fifty-two thousand exhibits. The royalty of Europe and the world attended the festivities, including Tsar Alexander II of Russia, fresh from freeing the serfs; the brother of the emperor of Japan, making a novel and much-discussed appearance after Japanese isolationism was shattered in the 1850s; Kaiser Wilhelm of Prussia and his minister Otto von Bismarck, their power growing after Prussia's defeat of the Austrians in 1866; Prince Metternich and his Austrian lord Franz Josef, whose arrival was delayed because of the execution of his brother Maximilian by a Mexican firing squad; dozens of lesser royalty from around the globe; and, most remarkably, the Ottoman sultan and the khedive of Egypt.

The Ottomans had tried to conquer Europe for centuries. The sultan had been feared as a secretive, omnipotent monarch guarded by eunuchs and retainers in his palace by the Bosporus, catered to by an army of servants and the slave girls of the harem. Now the sultan put on a fez and Western clothes and journeyed to Paris. He was, as the French discovered, a man much like any other.

Abdul Aziz understood that the balance of power had shifted permanently, and that the best he could now strive for was acceptance into the fraternity of European royalty. His hold over his Balkan provinces was tenuous, and becoming more so, but even though it was known as the "Sick Man of Europe," the Ottoman Empire still ruled the entirety of what later became Syria, Jordan, Iraq, Saudi Arabia, Israel, Lebanon, Kuwait, Armenia, Bulgaria, Serbia, Albania, Bosnia, Egypt, Libya, Tunisia, and, of course, Turkey. The sultan intended to reform the bureaucracy, rebuild the army, and improve the economy in order to participate as an equal in great-power diplomacy. He was young and adventurous, and Paris was the place to go for a wealthy monarch in the summer of 1867. The sultan was joined there by Ismail, who knew the city and combined his visit with a trip to England. Though Ismail attracted a fair share of attention, the sultan's appearance signified something more extraordinary. Abdul Aziz was the defender of Mecca and Medina, the holy cities of Islam. He took pride in serving as the protector of the pilgrims who made their annual hajj. But in 1867, he did not look to Mecca first. He looked to Paris, the metropolis of Europe, and it was there that he made a pilgrimage, in order to see and be seen.

He went, and Ismail went, to gaze on the industrial wonders of the day and to listen as Napoleon III declared that "representatives of science, the arts, and industry had raced [to Paris] from every corner of the corner of the earth." There were paintings to be seen, music at the Opéra by Offenbach and Rossini, and ecclesiastical decorative arts ranging from organ pipes to wax figurines of noted saints. There were livestock exhibits, botany displays, and agricultural specimens. There were restaurants and cafés and booths selling and preparing exotic foods. Gold and silver medals were awarded to honor the best exhibits, but everyone agreed that the stars of the exposition were the machines.

The German arms manufacturer Krupp assembled a massive fifty-ton cannon made of steel, with its equally massive shells lined up beside it. As children scampered around the weapon, their parents could not have imagined that three years later, similar guns would rain hell on the city. The Americans demonstrated their latest telegraph technology, and both Cyrus Field and Samuel Morse gave lectures. There was a display of Steinway pianos and McCormick reapers. The United States Sanitary Commission exhibited ambulances, and American entrepreneurs brought motorized apple-coring devices and primitive dishwashers. One French inventor presented a device that turned rabbit skins into felt hats. There were displays of guns, coins, perfumes, flowers, furniture, pottery, and clothing. The Ottomans provided a lush display of the riches of their empire, but though the sultan nominally ruled Egypt, at the Universal Exposition his vassal outshone him, mostly because of two small rooms.

Egypt presented several exhibits. In the main hall, there was a series of rooms decorated in the classical style of medieval Cairo, and filled with cotton, wool made in Alexandria, armor for horses and camels, musical instruments, reclining couches, Nubian tobacco, beans from Sudan, tamarind, lentils, peas, guano, sugar cane, pottery from Aswan, engraved knives, and decorative silks. In the park outside, the Egyptian pavilion was located in the "Oriental" quadrant, near the mini–Ottoman mosque with its minaret and its replica of a Turkish bath, and next to the faux palace of the bey of Tunis. The pavilion was designed by Auguste Mariette. The exterior of the building copied the ancient monuments of Egypt, and was fronted by a hundred-foot-tall plaster-cast copy of the Nile Temple of Philae (which was actually a hybrid of pharaonic and Hellenistic styles), complete with hieroglyphics on the pylon. The path to the entryway was lined with sphinxes. Inside, Mariette assembled a selection of recently discovered antiquities. Visitors walked past statues of Hathor the bull god, and then past Osiris and his sister and lover, Isis. There were sarcophagi, scarabs, jewels, gold, and five hundred mummified skulls. And then there were two small rooms that housed the Suez Canal.[3]

For many entrepreneurs, the festival was an opportunity to gain exposure and even patronage. The Canal Company was no different. It used the occasion to present its best face to the public just before it turned to that public for additional financing. If all went well, millions would visit the Egyptian pavilion and be amazed by the magic of the canal, and then, when the company announced toward the end of the exposition that it was seeking a hundred million more francs, it would be buoyed by the recent warm memories of the exhibit and by the out-pouring of patriotism that the festival had produced. The company would have mounted an exhibit whether or not it needed a loan, but obtaining the loan might have been far trickier without it.

The Suez exhibit was divided into two sections: an outer, rectangu-lar room, and an inner, circular one. The outer room was stuffed with drawings, topographical maps, city plans of Port Said and Ismailia, and photographs, as well as geological samples and models of the new dredgers and steam elevators and railroad tracks and jetties. The cen-terpiece was a large bas-relief map of the maritime canal and the Sweet Water Canal, with raised portions for the plateaus, depressions for Lakes Timsah and Balah and the Bitter Lakes, and miniature replicas of the machines.

In the inner room, the lighting was dim, and there were no objects. There was only a diorama, but that word doesn't capture the allure of the exhibit. Though primitive by the later standard of motion pictures, the diorama manipulated light and three-dimensional images to convey an illusion of watching the canal take shape in real time. Devised at great expense by the decorator of the new Opéra in Paris in collabora-tion with one of the company's architects, the diorama was a collage of painting, photographs, and figurines. It filled a semicircle, from the Mediterranean on the far left to the Red Sea on the far right. It depicted villages, machines, and cities along with the canal itself, from the harbor at the northern entrance to the lakes along the way. There were figurines of workers, dredgers, buildings, and ships. Visitors emerged from the exhibit exhilarated at the verisimilitude and claimed that they felt transported from the busy parks of Paris to the bustling Isthmus of Suez. In the words of Théophile Gautier, "After leaving the rotunda, it seemed as if one had actually made a voyage to the isthmus, and that one had passed from one sea to the other on one of those steamships that go from Marseille to Calcutta."[4]

The company made sure that a senior executive was always on hand, at least during the first months. Sometimes, Lesseps himself greeted visitors, especially when there were visiting dignitaries. One correspondent, awed by "this work of titans," and delighted by the site of "Ismailia, a European oasis in the desert of Egypt," was thrilled to be given a tour of the diorama by the secretary-general of the Suez Canal

Company. "No one could have done a better job," he subsequently wrote, "conveying the efforts, the difficulties overcome, the fears, and the defections, and the obstacles created by nature, or, even worse, by man. As we surveyed the works, we were able to appreciate this odyssey with one of the most energetic actors in this drama." The diorama also provided an excuse for hyperbolic musings on East and West, on the triumph of science over nature, and on the virtues of Ferdinand de Lesseps. The correspondent for the newspaper *Le Soleil* commented, "It is Egypt, forward sentinel of civilization, geographically placed as a link between the old Orient and the young Occident, that is in the front line of this peaceful struggle against the workings of nature. And it is because of the intelligence and the energetic will of one man that the project will be a success. This man, as all the world knows, is M. F. de Lesseps, who will achieve a work accomplished long ago by the pharaohs, destroyed by time, and then revived as an idea by Bonaparte." Such sentiments were commonplace, and others spoke of the canal as the perfect combination of individual spirit and the "light of science."[5]

Lesseps could not have asked for more favorable coverage. In the midst of a crowded festival, with thousands of exhibits vying for attention, the Suez Canal was given accolades and the diorama was lauded for its technical proficiency and aesthetic artistry. Two of the themes of the Universal Exposition were science and the Orient, but almost all of the exhibits concerned only one or the other. The Egyptian pavilion combined elements of both. Mechanical dredgers were juxtaposed with ancient gods; mummies lay near miniature mosques; and the Suez Canal diorama was installed inside a building that looked like a Ptolemaic temple. Not surprisingly, the Egyptian exhibits were among the most popular and praised aspects of the fair. Eugénie, an arbiter of what was fashionable in the Second Empire, was so taken with the display of a gold diadem and other jewels rumored to have belonged to the wife of Pharaoh Amenhotep that she tried to purchase them from Mariette. She promised him the title of senator-for-life and a stipend. He refused, and the French infatuation with all things Egyptian intensified.[6]

The exhibits made the subsequent task of raising the loan easier. When the company announced the subscription, it was treated respectfully by the increasingly independent French press, though several papers questioned the company's financial statements. The English press was also fair, except for *The Times,* which remained as scornful as ever. *The Standard* said that, "The Suez Canal is a grand and sublime undertaking," but the paper lamented that there was insufficient third-party information about the spiraling costs.

Other journals rhapsodized. "The stupendous character of the project effecting a maritime communication between the Mediterranean and the Red Sea grows more and more on the comprehension of the world," ran an English editorial. "All who have visited the works, or examined the splendid panoramic view and models of the cuttings, buildings, ports and towns exhibited by M. Lesseps at the Champ de Mars, must be satisfied that the work draws near its end, and that it is within the range of probability that eighteen months will witness its completion."[7] The diorama had had the desired effect. Unable to bring all interested investors and opinion-makers to the isthmus to see the canal for themselves, Lesseps and the company brought the canal to the center of Paris, where it could be viewed by all.

But the flurry of attention was not all for the best. For the first time since Napoleon became emperor, the Legislative Corps had begun to exercise real authority. No longer a rubber stamp for the emperor or for Morny (who had died in 1865), the legislature, led by Émile Ollivier, decided to address the issue of the Suez loan on the principle that the affairs of the company directly affected the affairs of France. Ollivier felt that the company was using questionable tactics to make the loan attractive to the public. After an initial offering that closed in September, many of the bonds hadn't been sold, and the company wanted to introduce a lottery that would offer prizes with the bonds. This controversial gimmick skirted the boundaries of laws from the 1830s that regulated loans. On Ollivier's initiative, the deputies engaged in a heated debate in 1868. Some wanted the company to disclose a more detailed account of its expenses. Others demanded that it provide a better accounting of its future revenue before it tapped French investors for more money. Echoing questions that had been raised in the press, a number of deputies declared that, although they supported the project, they wanted more transparency from Lesseps and the company.

Speeches were made, and questions were asked. It wasn't clear that the legislature had the authority to interfere. For the most part, the deputies did not threaten any legislative action and simply argued about whether the company should seek a public loan. Some members were appalled that the legislative body even bothered to discuss the issue. Viscount Lanjuinais rose to criticize the deputies for wasting time on the question. He claimed, reasonably enough, that the government had neither the power nor any good reason to become involved. The Suez Canal Company, he continued, was a publicly owned firm operating within its rights, and nothing in prior law or precedent suggested that the Legislative Corps possessed the right to tell such a company what it could and could not do in the financial markets. If it were an issue of changing the laws governing public loans, that would be a legitimate debate, but the particular doings of the Suez Canal Company were beyond their purview.

Another deputy seconded that concern. Jules Favre, an ardent man of the left who would soon rise to the fore of the Third Republic, asked plaintively if he had missed something. Was the question of the Canal Company and its loan a private endeavor, or had it, unbeknownst to him, suddenly become part of the national honor of France? His tone was facetious; the question was rhetorical. Yet Favre was one of the first to identify the changing status of the canal and its company. It had begun as an admirable, improbable dream. As it grew, it became the focus of international rivalries, and it then matured into a substantial but troubled engineering project of a large joint-stock company. And now, it had become the embodiment of French national honor. As one Parisian paper noted, "Public attention is now fixated on the piercing of the Suez Canal, and it is impossible not to consider it a political enterprise as well as an industrial and commercial endeavor."[8] Questions notwithstanding, the canal had the overwhelming support of the French public, and hence of the legislature. After several days of discussion, the Legislative Corps, and then the Senate, passed a resolution permitting the company to use a lottery with prizes in order to market the loan to the public.

The increasing centrality of the canal in French politics and culture was matched by the growing drift of the Second Empire and an emperor who was succumbing to age, ill health, and stymied ambitions. His adventures overseas had mostly been embarrassments, except for the costly consolidation of French rule in Algeria. His goals to dominate the politics of Western Europe had also fallen short, and he watched as Germany grew at Austria's expense. The economic expansion of the 1850s had given way to inflation and social unrest in the 1860s, and no matter how much he loosened his grip on power, France needed to do more than let off steam.

As Napoleon's star began to fade, Lesseps's shone ever more brightly. He had always walked among the elite, but he himself had not been in the first tier. The exposition and the loan subtly marked his graduation from entrepreneur to man of affairs, a noble in all but title, which he would acquire soon enough. With success, his ambition actually grew. He began to think about his impact on the future commerce and politics of the world. He accepted more invitations to lecture, and he basked in the applause and the toasts in his honor. Some observers noticed that he was not a man with an ego that could be sated. "The most remarkable man of this era is without question Ferdinand de Lesseps. The tireless and persevering promoter of the piercing of the Suez Canal has only one fault: it is to be so great that his shadow enthralls those who are enveloped by it." But however many noted this aspect of his personality, many more happily compared him to the pharaohs and placed him in the rarefied air of "superhumans" who change the course of history.[9]

And then there was Ismail. The Universal Exposition had a strange effect on the world's only khedive. Greeted by Napoleon and Eugénie at the Tuileries Palace, he was asked by the emperor what the word "khedive" meant. Before he could say anything, Nubar skillfully answered for him. "It is purely honorific, Sire, but the importance of the firman is that it makes Egypt completely autonomous and gives the khedive the right to make commercial agreements, including the right to negotiate customs rates." Nubar was only expressing what the khedive himself felt. That summer, in both France and England, he was greeted with great deference and wild enthusiasm. At public events, crowds gathered to cheer for him. He became a celebrity. His every move was covered in the press, and he was praised for his generosity, grace, and poise. He was not the only visiting dignitary to receive such scrutiny—the sultan was also a hit that summer—but he was acutely aware of his escalating fame. And, having observed the new Paris of Napoleon and Haussmann, and having talked to both of them about what they had done, he returned to Egypt more determined than ever to renovate his own cities and make Cairo into a modern metropolis that could hold its own against any capital in Europe.

Whereas he had once been cautious in his spending, Ismail now decided that the end was more important than the means, even if that required spending far more than he had. He authorized a new round of public works, including the transformation of Cairo and Alexandria.

These projects entailed more foreign loans, both public and private, at a time when cotton revenue was no longer increasing. The end of the Civil War in the United States led to the resumption of American cotton exports. In 1865 and 1866, that caused severe fluctuations in the price of cotton, and though these eventually stabilized, competition kept the price down. Egypt was able to maintain and even boost its revenues in the late 1860s by bringing more acreage under cultivation, but the cotton income was not sufficient to cover the costs of the public works. Ismail also stepped up Egypt's sugar production, and ordered the construction of costly refineries. That only led the county deeper into debt. Though some of his projects made long-term economic sense, trying to turn Cairo into a Nilotic version of Paris did not. Ismail hungered for the respect that the crowds had shown him that summer in France and England, but Egypt suffered because of his appetite.

Ismail needed to improve the Egyptian economy and revamp the government. Roads, canals, administrative reform, telegraphs, trains, and bridges—these were imperative if Egypt was to survive as an independent state in a world increasingly at the mercy of European imperial expansion. Palaces, gardens, wide boulevards, and the finery of the world, on the other hand, allowed Ismail to act like a member of the European royalty but did little for Egypt. He took on ever-larger loans at punishing interest rates in order to fund both the imperative and the frivolous. Given the annual revenues his treasury collected, there was only one way to repay these: the Suez Canal. Having tried and failed to gain control of the canal for himself in the first years of his reign, Ismail then linked his future to its success. As of the late 1860s, that seemed reasonable. The canal showed every sign of fulfilling its promise as a waterway that would alter the pattern of international commerce and enrich both Egypt and the company. Ismail was the largest shareholder and would profit handsomely once the shares started to yield dividends. Under the terms of the original concessions, the Egyptian state was also entitled to 15 percent of the net earnings, and the country stood to gain from customs duties as well.[10]

Having fought against Lesseps, Ismail now treated him as an ally. For Lesseps, however, the khedive was no longer a concern. In the space of a few years, Ismail had gone from a crucial determinant of the canal's outcome to one of many powerful and influential people vying for a prominent role in its future. Lesseps did not snub Ismail, but he did not go out of his way to cultivate him either. Instead, he attended to more pressing issues. Work on the canal was far from over. The cities of the isthmus were burgeoning, and Lesseps was becoming the de-facto monarch of a small realm. The company had to set rates for passage and then convince shippers that the route was more viable than the existing alternatives. And though the completion date was almost two years away, Lesseps was already envisioning the inaugural ceremony. The project had begun quietly in Said's tent; it would end with a festival to rival the Parisian exposition.

# CHAPTER EIGHTEEN

# THE FINAL STAGES

A T THE END of 1867, less than half of the excavation was finished. A maritime canal existed between Port Said and Lake Timsah, and flat-bottomed ships and barges regularly made the transit. But the channel was not nearly deep or wide enough to accommodate larger vessels, and millions of cubic meters of earth still had to be dredged. The average final width was supposed to be three hundred feet at the top, though in certain spots it would be as narrow as two hundred feet, while the minimum bottom width would be seventy-two feet. The depth would also vary, though the goal was to have a minimum of twenty-six feet, which would be just enough to allow passage for the latest steamships. The northern half of the canal was almost complete, but in many spots the depth was only ten feet. South of Lake Timsah and Ismailia, the work was even less advanced. The six miles of the Serapeum ridge remained untouched and kept the waters of Lake Timsah from flowing into the dry bed of the Bitter Lakes, while the rocky Chalufa ridge still stood between the Bitter Lakes and Suez. The Bitter Lakes themselves required little excavation or preparatory work. They were below the level of the canal, and the plan was to inundate them once the two ridges on either side were pierced. But the dry lakebeds posed dangers to the workers. They were dotted with large crevices that appeared unexpectedly and could kill camels and horses. While the work on the jetties of Port Said was proceeding smoothly, tons of cement still had to be mixed, poured into molds, baked in the sun, and then put in place.[1]

These logistical problems did not threaten the actual completion. Voisin, the Dussauds, Borel, and Lavalley had solved the technical issues, but now they confronted a different challenge: time. The completion date was set for October 1869. The company guaranteed its shareholders that the work would be finished by then, and it claimed that if the various contractors failed to adhere to that schedule they would be fined as much as five hundred thousand francs for each month of delay. As he had a decade earlier, when trying to gain support for the nascent idea, Lesseps suddenly seemed to be everywhere. He went from city to city, and spoke about the progress of the canal. He was asked to represent Marseille in the Legislative Corps. Lacking the time but honored by the request, he reluctantly agreed and then lost the election. The fact that he was asked, however, was a sign of how illustrious he had become.

At event after event, he appeared confident and relaxed, yet there was a race going on, one that he would have preferred not to run. The company was spending money at an astonishing pace. The completion date was not the product of a careful calculation of how long the remaining work would take; it was when the company would exhaust its funds. The work schedule was calibrated to the rate at which these funds were diminishing. The shares were down, trading more than 20 percent below par value, at about four hundred francs. The loan was taking longer than anticipated to be finalized. Lesseps once again was part juggler and part confidence man. He presented a bold face, with the sweat never visible. He had faith that the bulk of the work would be completed before the money ran out, but it was no sure thing.

As the work sped up, the cities along the canal grew. People arrived daily, some to operate the expanding number of machines, some for the tedious digging between Chalufa and Suez. They came from across Egypt and the Near East. Bedouin tribes arrived, led by their shaikhs, looking to be hired. Groups of men from Nubia crossed the hundreds of miles of desert to find work. Thousands more came from Europe, drawn by the prospect of a new life and quick wealth. In 1865, the canal zone had a population of ten thousand; in 1866, that grew to eighteen thousand; in 1867, twenty-six thousand; and by 1868, thirty-four thousand, almost evenly divided between Europeans and non-Europeans. By 1868, Port Said and Ismailia each had a population of about ten thousand. Only a few years before, nearly everyone in the canal zone had been employed by the company, but many of the new arrivals didn't work on the canal at all. Instead, they set up businesses or found jobs as servants, storekeepers, clerks, lawyers, gardeners, cooks, blacksmiths, and telegraph operators. They became merchants or agents for merchants and competed for an early advantage in what promised to be a lucrative future. The cities took on their own life, inextricably bound to the canal and the company, but with their own urban rhythms and their own municipal needs.

The inhabitants were all immigrants, and they spoke a variety of languages. The canal zone became known as a latter-day "Tower of Babel." One observer joked that in Port Said "you speak bad Italian to the Arabs, even worse Greek to the French, and an impossible Arabic to the Dalmatians."[2] Lesseps boasted that the ethnic mix of the isthmus lived in harmony, freedom, and liberty. The company made the region sound as if it embodied the precepts of the French and American Revolutions, and that in the canal zone people could pursue their dreams unfettered by unjust laws. That was partly true. Much like the American West in these same years, or the pampas of Argentina, Suez was a raw area that hadn't yet developed the layers of established society. But with that freedom came a good deal of conflict, and violence.

In 1868, the company began another struggle with the Egyptian government over customs duties and jurisdiction. The canal zone was in many ways lawless. Neither the company nor the government had been able to keep up with the population growth. Though the company was superbly organized, it didn't have the staff to run small cities. Egyptian officials administered the native populations of the isthmus, but they were sent from Cairo and depended on the good will of the company for such basic needs as telegraph communication, food, and supplies. The government, naturally enough, contended that it alone had the authority to adjudicate disputes and punish those who broke the law. That raised a crucial question: what laws applied to whom? If a Greek canal-worker stole from a Serb, would he be tried in an Egyptian court? Would he be brought in front of a company disciplinary committee and subjected to the Napoleonic Code of France? Or would his own consul handle the dispute, as consuls had done in the Ottoman Empire for centuries under the system of capitulations?

Scornful of "Oriental justice," European governments wanted simply to extend the old system of capitulations and establish legal structures that would prevent foreign nationals from being tried in Egyptian courts. In the past, however, the capitulations had only covered the few hundred foreigners living in Alexandria. With the influx of thousands to the isthmus, the same system would have made a mockery of Ismail's claim to govern all of Egypt. If a majority of the people living in the canal zone were foreign-born, and if none of them were subject to Ismail's authority, then the government would have minimal control over the area. It had only the right to police and tax Egyptians living there, and not even that when an Egyptian had a dispute with a European. There was no way Ismail could accept the de-facto loss of sovereignty over the canal zone, even if he personally had a large stake in the company.[3]

As Ismail understood, a state—especially an autocratic state—maintains control with armies and courts. A government that cannot create and enforce laws is no government at all. Ismail fought for the right to determine the laws of Egypt and to enforce them, but the very fact that he had to fight at all showed just how tenuous his independence had become. Having gained autonomy for Egypt within the Ottoman Empire, Ismail immediately found that autonomy compromised by a hungry and expanding Europe, with the Suez Canal Company in the vanguard. Contrary to the hopes of Said and the promises of Lesseps, the closer the canal came to completion, the further the dream of an Egyptian renaissance receded. Lesseps could promise a renewal of Egypt, but he could not change European attitudes that saw Egypt as backward. One journal stated that consular courts should be the only system of justice, one that Ismail ought to embrace because such courts would be "an excellent school for the Orientals." Ismail and Nubar could legitimately argue that Egyptians, Arabs, Ottomans, and Muslims in general had a longer, more sophisticated tradition of jurisprudence than any European state, but in a world increasingly dominated by the power of Europe, such arguments carried little weight. More common was the belief that the canal was reversing a pattern established by Islam. "In the 14th century," said one Catholic journal, "Muslim fanaticism closed off commerce between Europe and Asia, which forced the peoples of Europe to discover new routes to India. In the 19th century, thanks to a most extraordinary revolution of ideas and a most amazing enterprise, the pashas of Egypt reopened the doors of the East to the West." Though the legacy of Muhammad Ali was honored, centuries of Arab and Muslim culture were denigrated and distorted.[4]

The question of jurisdiction was not resolved until the middle of the next decade, and then only after years of complicated negotiations led primarily by Nubar Pasha. In the end, a new system of "mixed tribunals" was established that replaced the capitulations with hybrid

courts, composed of both European and Egyptian judges, to try disputes between different nationalities.

The company took an interest in the capitulations, but it did not
truly care who had jurisdiction over a Bulgarian accused of a crime in
Port Said. It did, however, care greatly about customs duties. Lesseps
had always promised that the canal would be a neutral waterway,
accessible to all navies, but Ismail expected to collect taxes on goods
that entered or left Egypt via the canal. Goods transported through the
canal on their way, for example, from Marseille to Calcutta, would not
be taxed, because they would not technically be imported into Egypt.
But the company and the government would still profit from the fees
charged to use the canal.

The logistical arrangements between the company, which would
operate the canal, and the government, which would collect duties,
had never been mapped out. Until the canal was near completion,
there had been no need to. In 1868, however, much to the government's
surprise and dismay, Lesseps broke a long-standing tacit agreement
and claimed that the canal should be a duty-free zone. Writing to
Ismail in April, he said that, because the government would collect 15
percent of the net profits of the canal, no goods imported and consumed in the canal zone should be subject to customs. It was an
apples-and-oranges argument born of hubris rather than reason. The 15
percent reserved for the government had nothing to do with customs
on goods imported into Egypt at Port Said, or Ismailia, or Suez. Lesseps
also said that the original concession, by granting the company the
right to import construction materials for the canal duty-free, extended
to goods consumed by the inhabitants of the towns along the canal.
Not surprisingly, Ismail disagreed.

The company and the Egyptians took their dispute to an arbitration
panel composed of consuls. In 1869, the panel ruled that the government did have the right to collect taxes, but that the company should
have wide latitude to import materials, classify them as construction-
related, and not pay any duties. This was not an issue that would go
away, nor was it one that the company and the government were able to
resolve conclusively before the canal opened. But by even suggesting
that the canal zone should be duty-free, the company put the Egyptian
government and the world on notice that it thought of itself as much
more than a commercial venture.[5]

In the eighteenth century, the British and Dutch had chartered semi-sovereign entities to undertake commercial exploration. Firms like the East India Company were chartered by the crown, set up their own armies, and then conquered and governed territory. In similar fashion, the Universal Company of the Maritime Canal of Suez had grander ambitions than simply ferrying ships from the Mediterranean to the Red Sea and paying dividends to its shareholders.

Though it was now a vast and complex organization, the company retained the indelible mark of its founder. From the positions he took, the letters he wrote, and the speeches he gave, Lesseps seemed to have the following interpretation of what was right and just: With the exception of a few Bedouin tribes who sometimes paid taxes to the Egyptian government and often did not, the isthmus had been a vacant desert before 1860. Now lands were cultivated, and recently settled towns were thriving, and there was a contest over who would administer the region. Who should it be? On the one hand, the canal was within the territory of Egypt. On the other hand, the company had built the canal and the towns, and it was the sole economic attraction for the tens of thousands who had relocated to the isthmus. Lesseps had made all of it possible, and though others might have shared his dream, only he had been able to translate that vision into reality. Why shouldn't his company administer the isthmus, in addition to building and running the canal? The company had as much right to govern the area as an Egyptian monarch who had paid the region scant attention until recently. Even without direct jurisdiction, Ismail would make far more money than if the canal had never been created. So why, Lesseps reasoned, should the khedive object? Why shouldn't the Egyptian government be content with the fact that the canal would make Egypt more wealthy, more powerful, and more significant in world affairs than it had been for centuries?

Until the canal was actually opened, however, Lesseps did not press the point. Instead, he compromised with the khedive to ensure that payments continued to flow. The company faced a cash squeeze throughout 1868 and 1869, and the khedive was a reliable source of emergency funds. Ismail might have extracted more concessions in return, but his attitude at this juncture, like that of Lesseps, the company's directors and shareholders, and most who were involved, was simply to get the thing finished.

One of the remaining decisions was what to charge for passage. In 1867, the company had already begun to set rates. It had also started to transport goods from Port Said to Suez, but it was a makeshift system. The canal between Port Said and Ismailia could accommodate mid-size barges, and from there, smaller barges could float down the Sweet Water Canal, which ran roughly parallel to the line of the maritime canal. Though ships could not yet pass through the isthmus, this was still a significant improvement over any other trans-shipment of goods from Europe to Asia via Egypt. Over the next three years, the route generated several million francs in annual revenue, but that was a fraction of what the actual maritime canal could produce.

In May 1867, the company announced its preliminary fee structure. Depending on the type of cargo, ships would be charged between twenty and twenty-five francs per ton. Annually, about ten million tons of goods were shipped between Europe, Asia, and America around the Cape of Good Hope. The company based its calculations on the assumption that half of that would soon pass through the canal. As it turned out, the company aimed too high, and in response to complaints, prices were lowered to as little as ten francs per ton (which was actually the rate that Lesseps had initially planned in 1856) and ten francs per passenger, which would mean approximately sixty million francs a year in revenue. Unfortunately, these predictions of initial use substantially overestimated how quickly world trade would adjust to the opening of a new route.[6]

Aware that simply building the canal would not guarantee that people would use it, Lesseps and the company began an extensive public-relations campaign. There was, as yet, no established industry for promoting companies and ventures, but the principle, then as now, was the same: get people excited, make them enthralled, and spread the word.

To begin with, the company installed the diorama from the Universal Exposition as a permanent exhibit near an entrance to the newly opened Bois de Boulogne in Paris, and a painter was hired to keep it up to date. Lesseps himself stopped by the exhibit when he was in Paris, surprising visitors with an impromptu update on the remaining work.[7] He and other company officials took every opportunity to link French national honor to the canal's success. Though this would do little to induce English merchants to use the canal, it would at least solidify the company's place in France. If the anticipated revenues did not materialize as quickly as hoped, the company would need the generosity of the French bourgeoisie if they were called on for additional loans. In order to inspire enthusiasm in England, the company tried to generate favorable press. Journalists always need new stories, and several high-profile visits, one by Lord Mayo at the end of 1868, and the other by the prince and princess of Wales, gave them a reason to write about the canal.

Lord Mayo was interested in the canal because he had been appointed viceroy of India, and was on his way to assume his duties. Much to the surprise and delight of company officials, *The Times* wrote glowingly of the visit. The correspondent marveled over how much had changed in a few short years. The population of Port Said was nearly twenty thousand; its harbor teemed with frigates, barges, steamers, and sloops. Kantara, once a dry patch at the edge of the Lake Manzala swamp, boasted a hospital, tidy streets, and, by all accounts, a superb dinner at the Buffet de la Gare. In the words of the illustrator Édouard Riou, "The chicken was perhaps a little fleshy but the cutlets were succulent and the wine good." In case the chicken failed to entice visitors, company officials emphasized that the settlement was the midpoint of the ancient route from the Holy Land to the Nile.

*The Times* expressed concern about the inundation of the Bitter Lakes, which was scheduled to begin shortly, as soon as the dam that kept back the waters of Lake Timsah was removed. At some point several months in the future, the inundation would be completed with the removal of another dam, to the south, allowing the waters of the Red Sea to merge with those of the Mediterranean in the Bitter Lakes. No one knew what would happen when the two seas merged. Would the water level rise above expectations? Would there be excessive pressure, or too much volume for the embankments of the canal? It was also not clear how many months it would take for the lake beds, which extended nearly twenty-five miles and were as much as eight miles wide, to fill up. Other than raising this question, however, the normally hostile *Times* printed a series of laudatory articles. The canal, the paper declared, far from a threat to England, was "simply a waterway between two great seas—a shortcut from the Western to the Eastern world, and that is all." Whatever its financial prospects, it was a technical accomplishment that everyone would welcome. Lord Mayo claimed to be impressed and endorsed the canal as a benefit to English trade and to the health of the British Empire. Given that he would be assassinated in India three years later by an irate Afghani prisoner, he might have preferred to skip the trip.[8]

The canal received more positive publicity in the spring of 1869, when the prince of Wales, the future Edward VII, made a tour of the isthmus. That visit followed on the heels of the first official inspection of the works by the khedive, who had stayed in his chalet in the city named after him, and enjoyed the comforts that he was accustomed to in Cairo, Alexandria, or Paris. Lesseps timed the khedive's arrival with the inundation of the Bitter Lakes. When the khedive gave the signal, the dam was broken and the water began to pour in. Lesseps delivered yet another pitch-perfect encomium: "Moses ordered the waters of the sea to withdraw and they obeyed him; on your order, they will return to their bed." Four million cubic meters of water cascaded into the lakes in less than twelve hours.

The khedive, for his part, tried to make amends with Lesseps. "You know," he wrote in a letter designed to lessen the friction between the two, "that I have always had the most passionate sympathy for the great work that you have undertaken to join the two seas. . . . In visiting your works, I could only be convinced of their importance. . . . Today, no one can doubt that here, in several months, the Isthmus of Suez will be open to navigation and a new route will be opened to commerce and civilization. No one appreciates better than I the advantages, and as sovereign of Egypt, I can only be happy and proud that a part of the territory that I govern will become once again, as it had been long ago, the meeting point of Europe and the Orient." He vowed that, though pressing questions remained about customs duties and tribunals, he would do his part to settle such differences with mutual respect. But Ismail offered more than kind words. He agreed to purchase the telegraph systems then owned by the company, along with the quarries of Mex, hospitals along the isthmus, a number of stores, and fishing rights, thereby ending a dispute that had begun with a few carp in Lake Manzala. The total sum was thirty million francs.[9]

Ismail entertained the prince of Wales in Cairo before the English heir departed for the isthmus, and he provided his personal railway car for the prince's transport to the city of Suez. Lesseps organized a schedule of banquets and sightseeing, and the prince was treated to a tour of the works by Borel and Lavalley. The royal party stood on a wooden platform above the Bitter Lakes and watched as the waters slowly filled up the desolate expanse. Then they were conveyed by a company steamer to Ismailia—"the Venice of the Desert," according to the *Times* reporter. There, they stayed in the chalet recently vacated by the khedive, still staffed by "gold-laced and scarlet uniformed servants, with cooks and stores of all kinds . . . and a regiment of infantry with its band." A dinner attended by the luminaries of the company was served on the porch under the stars. A band played waltzes and an impromptu chorus of bagpipes while the prince, Lesseps, Voisin, and others "smoked the diamond and ruby studded chibouques, brought down by the corps of chibouqjees from Cairo." The next day, they went on to Port Said, passing the dredgers at work on the banks. At the mouth of the canal, the royal party was met by cheering crowds, a chorus of "God Save the Queen," and the strains of the "Viceroy's Hymn." The prince stayed on the khedive's barge, the *Mahroussah,* where dinner was served in a salon decorated with damask cloth and silver columns and lit by hundreds of candles. Impressed by what he had seen, the future Edward VII concluded that, in opposing the canal, Lord Palmerston "had been guilty of a lamentable lack of foresight."[10]

Another visitor to the canal attracted much less attention, an idealistic thirty-four-year old French sculptor opposed to the autocracy of the Second Empire and entranced by the Egyptian past. He had made a pilgrimage to the Holy Land and to Egypt when he was barely twenty-one, and as it had so many other travelers, the trip affected him deeply. He sketched the fallen statues that lay in the sand and spent days inspecting the half-excavated and partly restored Sphinx at the base of the Great Pyramids of Giza. He wrote about the emotion he felt looking at "centuries-old granite beings, in their imperturbable majesty . . . whose kindly and impassible glances seem to disregard the present and to be fixed upon the unlimited future." He did not forget what he had seen, and a decade later he returned to Egypt to make his own mark.

He had heard that Ferdinand de Lesseps planned to build a larger lighthouse at Port Said, and he lobbied for the commission to design it. He conceived of a statue, ninety feet high, guarding the entrance. One of the ancient wonders of the world had been the Colossus at Rhodes, built in the third century B.C., which observers described as a statue so immense that it straddled the mouth of the harbor, and ships would pass between its legs. No trace of it remained, but in the nineteenth century, as educated Europeans read the Latin and Greek texts of Strabo and Pliny, the memory of it was tantalizing. The young sculptor drew up plans for a towering statue of a female fellah. She would be draped in the robes worn by Egyptian peasants. To fulfill tne functions of a lighthouse, she would carry a torch, and her name would be *Egypt Bringing Light to Asia.*

Frédéric-Auguste Bartholdi was certain that Lesseps would embrace the idea. To his great disappointment, both Lesseps and the khedive declined. The statue was simply too expensive. Bartholdi dedicated himself to finding another location, and in 1871 he left for the United States. Fifteen years later, in 1886, a colossal statue was unveiled in New York Harbor, to commemorate the centennial of the Declaration of Independence. It was called *Liberty Enlightening the World,* more popularly known as the Statue of Liberty, and it was dedicated by an official delegation sent from France, one of whose members was Ferdinand de Lesseps.[11]

Though Lesseps passed on the ninety-foot statue, in order to generate excitement, the company exploited an unfolding drama: Would the work be completed in time? With assembly-line precision, the Dussaud brothers had been able to deposit thirty to forty of those twenty-ton blocks a day in the waters off of Port Said, and the jetties were finished in early 1869. The excavation, however, was not. In fact, in August 1868, nearly twenty-seven million cubic meters still had to be removed, almost a third of the total. Only thirteen months remained until the official deadline. The canal had always been perceived as a triumph of human ingenuity over the challenges of the natural world, but now it was portrayed as a sprint to the end. The suspense was whether the machines of Borel and Lavalley would be able to dig enough earth within the limited amount of time. The engineers themselves were skeptical, and worried that they would fail. The drama was real. All the company had to do was draw attention to it.

Correspondents were dispatched by papers throughout Europe to report on the stages of the work, and the monthly statistics published in *L'Isthme de Suez* were widely excerpted.[12] The dredgers operated sixteen or seventeen hours a day, and were running at full capacity. Adding to the suspense was the risk that too many of the machines would break and thereby make the promised date impossible to achieve. Since 1863, the work had relied on machines, but in 1869 the company hired thousands of unskilled laborers to dig by hand the soft sand on a ten-mile stretch between the Chalufa ridge and the port of Suez. Though no one could say for sure that the goal would be accomplished, Lesseps proceeded under the assumption that it would be, and that this final difficulty would somehow be surmounted, as every previous one had been.

With the date fixed, the company and the Egyptian government planned for a grand festival to celebrate the opening. In November, the most powerful and influential people in the world would be invited to Egypt to witness a historic event, and they could assess for themselves whether or not the canal was a practical, functional alternative to the long route to Asia around the Cape of Good Hope. After years of florid speeches and evocative dreams, people wanted to know one thing: could a modern steamship, fully loaded, make the passage from Port Said to the Gulf of Suez? If the elites of Europe, America, and Asia arrived and the canal was not finished, they would question whether it would ever be, and they would wonder if the company and Lesseps were up to the task of operating it. At best, they would go back to their homes annoyed at the pointless journey. At worst, they would never again trust what the company said, and Lesseps would be left with a reputation as a trickster who had saddled thousands of investors and two rulers of Egypt with a ditch to nowhere.

# CHAPTER NINETEEN

# THE DESERT IS PARTED

I T WAS A difficult summer for Khedive Ismail. His relationship with the Ottomans had deteriorated again, and the grand vizier, Ali Pasha, was attempting to dictate the size of both Ismail's army and the Egyptian state budget. This tug of war between Egypt and the Porte had been going on for decades. The two states had achieved an uneasy status quo, but each time a new crisis arose, officials in Cairo and Constantinople girded themselves for a serious confrontation. In hindsight, it is easy to dismiss these tempests as insignificant. At the time, Ismail, Nubar, the sultan, and the vizier wondered if the next step might be war.

While this diplomatic crisis unfolded in a series of tense meetings and cool telegrams, Ismail prepared for the inaugural ceremonies. The opening of the canal presented him with a unique opportunity to demonstrate to the powers of Europe how far Egypt had come under his stewardship. There would always be state visits. Ismail could look forward to many opportunities to host a prince or a queen, an emperor or a king, but singly, or in small groups, perhaps a few times a year, and rarely making a substantial impact either on Egypt or on Europe. But the completion of the canal was a singular event, and just as the Parisian exposition of 1867 gave Napoleon III the chance to strut for the assembled nobility of the continent, the canal provided an excuse for Ismail to showcase Alexandria, Cairo, and the cities of the isthmus.

The goal was to prove that Egypt had arrived. Ismail had spent millions on bribes in Constantinople to secure his title and hereditary rule, and he had spent millions more on canals, railroads, and the transformation of Cairo from a medieval Muslim city into a Middle Eastern Paris. None of that would do him or Egypt much good if no

one came. Having invested so much to transform the country, he naturally spent whatever was necessary to draw people to Egypt to witness
what he had done.

The company's goals were similar. No matter how much it publicized the canal through special brochures, press junkets, and advertising, the best way to demonstrate to the world that the canal was an
attractive alternative to the Cape of Good Hope was to get as many
people as possible to see for themselves what the company had
achieved. Company officials in Europe faced skepticism, and as they
began to set rates and look for business, they were confronted with
people who doubted that the work was nearing completion, and who
questioned whether it would be safe for passage even if the excavation
itself was finished. Years of negative publicity surrounding the feasibility of the canal, especially in England, had left far too many people
mistrustful of the endeavor. They expected the company to proclaim
the virtues of the new route, but they did not fully believe what they
heard. Until they had proof that the journey was not dangerous, that
the sands of the desert would not degrade the embankments, that the
shallow tides of the Mediterranean would not silt up the harbor at Port
Said, and that the currents of the Red Sea were easily navigable, merchants and shippers and navies would prefer the old route that they
knew, however long and inconvenient.

The company and the khedive, therefore, budgeted a million francs
to bring a thousand guests on an all-expense-paid tour of the canal for
its official inauguration.[1] The Egyptian government covered most of
the bill, largely because the company had barely enough cash to complete the work in time for the festival. At some point during the summer, the inauguration date was set for November 17. On that day, a
flotilla would gather at Port Said and begin a three-day journey through
the canal. There would be festivities at each of the canal cities, and
then a final round of celebrations in Cairo. Though the list of attendees
was still evolving, one person was confirmed: the Empress Eugénie,
the cousin of Lesseps, benefactor of the canal, and the guest of honor
for its opening.

Though she had supported Lesseps at several key junctures, she
had done so purely out of fondness for him. Egypt was an abstraction,
and the canal was simply a business venture. But during the Universal
Exposition, that changed. Eugénie was fascinated by the Egyptian
pavilion on the Champ-de-Mars, and the artifacts of ancient Egypt

caught her imagination. She had also been taken with the khedive, and the two had, according to the gossips, developed a close bond. In 1869, she was forty-three years old; Napoleon III was in his sixties. He had never been a faithful or attentive husband, and now his health was failing. After years of balls and parties, after decorating new palaces, overseeing the remodeling of old ones, and raising a young son, Eugénie was bored and curious.

For her as for so many other Europeans, Egypt and the Near East offered the chance of escape. For all the purported wonders of science and industry, there were strong currents of ennui and despair in mid-nineteenth-century European culture. The very people who were supposed to enjoy the fruits of the Second Empire or the benefits of the British Raj were often the same who yearned to escape to a distant past or to a remote desert. Eugénie began to look east just when her empire started to crumble. The trip to Egypt gave her a chance to see the ruins of the Nile and to be honored in her own right, separate from her husband, who was not inclined to make the trip. She would take her yacht, *L'Aigle,* and stop at Venice, Constantinople, and then Alexandria. From there, she would go to Cairo and set off for a two-week tour of the monuments of Upper Egypt, at the end of which she would return north, board her yacht, and arrive at Port Said by November 16.

That summer in Paris, the air buzzed with talk of the Suez Canal and the impending departure of the empress. Suddenly an invitation to the festivities became a precious commodity. Society women wrote to one another trying to find out who was to be invited. Egyptian fashion was the theme of the season, and jewelers developed copies of Egyptian necklaces. A theater impresario created a spectacle called *All of Paris Is Going to Suez,* with the subtitle *An Egyptian Fantasy in Two Acts.* The play began with the seated Sphinx talking with a mummy who is revealed to be Cleopatra. The second act had the statue of Memnon praising the modern marvel of the canal. "Today," Memnon announced to the audience, "the inauguration of the Suez Canal will occur. Millions of invitations have been sent through the populations of the universe in ebullition. They are coming, they are coming. . . . And the Egyptian desert will not long from now be invaded by a multitude of visitors. . . ."[2]

In anticipation of the opening, papers ran articles on the climate and vegetation of Egypt. There were reports that examined the growing French empire in Cochin China, trade with Japan, the possible effects of the canal on the port of Marseille, and the potential of increased tension with Britain. The shares of the company enjoyed a brief rally above five hundred francs. Entrepreneurs organized package tours for the inauguration. "Excursion to Suez: Travel Comfortably to Attend the Opening of the Canal and Visit the Isthmus!" For fifteen hundred francs, you could get a berth on a steamship outfitted with salons, loges, and even a piano, leaving Marseille on October 28, with medical service included and meals catered by the "best restaurateurs in Marseille." Commemorative picture books filled with artful illustrations of the canal were published. Lesseps himself held several dinner parties at his home on the Rue Richepance, near the Place Vendôme, where he entertained senators, deputies, and journalists and urged them to take a tour of the Nile and the canal works—paid for by the company—in the weeks before the inauguration. Though the khedive invited a thousand official guests, the company also arranged special tours for merchants, illustrators, artists, sculptors, painters, essayists, heads of chambers of commerce, and members of the international press. Most of them leapt at the chance to see Egypt and the canal, but Lesseps wooed them nonetheless and promised that the opening ceremonies would be suffused with the magic of the Orient and would surpass anything in *A Thousand and One Nights*.

The English, though ambivalent, were as intrigued as the French. Some observers, recognizing that the canal would probably be up and running within months, assailed the long-standing policy of the British government to oppose the project. "The Suez Canal," wrote the *Daily Telegraph* on August 26, "should have been constructed with English capital and by English energy; and we, as a nation, have little reason to thank the diplomatic authorities and the scientific experts who kept on assuring us, year after year, that the scheme was an absurdity. . . . If M. de Lesseps had not been a man of the stuff and stamp of which all great inventors are made, if he had not toiled on to the attainment of his end in spite of every hindrance, the Suez Canal would now exist only on paper. . . . The recollection of the period when our public men pooh-poohed the very notion that the canal could ever be anything other than a colossal folly ought surely make us somewhat sceptical of the assertion still frequently made, that when the canal is completed it can be of no practical use. . . . The opening of the new water highway between the East and the West will mark an era in the annals of humanity; and we shall be glad to find the French nation, in the person of Empress Eugénie, doing honour to the undertaking which will add no mean trophy to the glories of France and the British Empire." Even if the lure of the impending inaugural festival was not as strong in London as in Paris, invitations were still coveted. Thomas Cook, who was fast developing the modern tourist industry, unveiled several packages that combined the wonders of ancient Egypt with a sightseeing tour of the newest marvel, the canal.

As the summer came to an end, Lesseps finalized his guest list. He was disappointed, though not surprised, when the sultan turned down an invitation to attend. The Porte had never warmed to the canal, and with relations between Cairo and Constantinople strained, Abdul Aziz did not wish to honor the khedive. There were other refusals, but there were far more acceptances. The emperor of Austro-Hungary, Franz Josef, said that he would be happy to attend, in his own ship. The heir to the Prussian throne said yes, as did the prince of the Netherlands. Lesseps was involved in every stage of the planning for the festival, from deciding which ships would go in what order to which clerics would give the benediction at Port Said to the design of the company's pavilions and tents.

In August, he hosted the annual meeting of the general assembly of shareholders. Though he remained as popular as ever and was treated to several standing ovations, he was required to present a tally of expenses. It was common knowledge that the project had exceeded its original two-hundred-million-franc budget. Lesseps announced that to date 404 million francs had been spent, and the work was not yet complete. Tens of millions more were needed in order to pay for the removal of the last cubic meters and to meet obligations on the next installments of interest on the loan. Borel and Lavalley would be subjected to a large fine if they did not finish in time, but would be owed a bonus of several million francs if they finished early. Lighthouses had to be built at Port Said in time for the opening. Though not as costly as the monument conceived by Bartholdi, they were not inexpensive. And though the tone was triumphant, buried in the report was a potentially embarrassing fact: even if the canal opened on schedule, some work would need to be done in the months after.[3]

In the middle of August, there was a brief ceremony to commemorate the last major stage of the construction. The Bitter Lakes had nearly filled up with the waters of the Mediterranean via Lake Timsah; the Chalufa ridge had been cleared; and the final miles to the city of

Suez had been dug by hand. What remained was a final dike preventing the Red Sea from flowing north. When that dike was broken on August 15, the seas flowed quietly and peacefully together, putting to rest the age-old fears that something terrible would happen when the waters mingled. Voisin Bey, who presided over the ceremony, wrote to Lesseps, "The appearance of the canal was splendid. I regret that you were not there to enjoy it with us. The current was barely discernible; and the embankments were just as they should be."[4]

That should have been the last dramatic event in the actual construction, but workers then discovered a huge block of stone in the channel that cut through Chalufa. Soundings of the channel's depth revealed that the stone was close enough to the surface to prevent ships from passing through. The rock was too hard for dredgers, even if they could have been adapted to the correct angles and depth. The crisis was even more severe because Paul Borel died suddenly in mid-October. Distressed, Lavalley returned to France and was not able to lend his expertise to solving the problem. Lesseps asked the khedive for help, and they arranged for an emergency transport of gunpowder. Just before the flotilla was scheduled to leave Port Said, workers embedded powder in the hundred yards of rock and detonated it.

In the meantime, people had started to land at Alexandria. To prepare Cairo for their arrival, Ismail had used thousands of workers, and whether or not it was technically termed a corvée, it is a fair assumption that many of them would have preferred not to spend frantic weeks putting the final touches on a new guest palace in Gezirah, decorated in a hybrid style that combined the gilded excesses of the Tuileries with Moorish decorations more reminiscent of the Alhambra than of any building in Cairo. Workers were also dispatched to finish the Cairo Opera House in time for its November 1 premiere. The Ezbekkiyah Gardens, which were a perfect copy of a European park, were trimmed and decorated. Avenues were swept, and the Qasr al-Nil Palace was given a thorough sprucing. The streets and public buildings of Port Said were also made presentable, and the khedive's chalet on the northern outskirts of Ismailia was renovated.

Eugénie's yacht reached Alexandria on October 22. The city struck her, as it did most Europeans who arrived there that fall, as unimpressive. Pompey's Pillar, Cleopatra's Needle, and a few other scattered Roman ruins were historically significant but physically bland, and the city itself was dedicated to commerce. Even with the Ras al-Tin Palace and the ruined fort in the harbor, it was too similar to a European port to engender much gawking from European tourists. Eugénie left immediately for Cairo, which she found far more delightful. She wrote to her husband that the city reminded her of Madrid. "The dances, the cuisine, and the music are identical," she said, though that probably owed more to the fact that she was staying in the new palace specifically designed for her by Ismail to evoke the beauty of medieval Spain. She did witness one exoticism on her first night—a performance staged by the dervishes. Already, the Sufi practice of ecstatic prayer was becoming a tourist attraction for Westerners. The empress also confessed to being slightly scandalized by the harem dances that were performed for her. Yet, in describing the belly dances, she sounded more like a blushing teenager than an offended matron. The next day, Auguste Mariette escorted her through the museum, but though he tried to give her long lectures, she constantly interrupted him with questions. The day after that, she departed for Upper Egypt on a Nile barge; she described to Napoleon how odd it was to see the "grandiose temples of the pharaohs" juxtaposed with the "semi-naked villagers living in miserable huts."[5]

Her trip coincided with an expedition organized by the company. A select group of journalists, artists, scientists, and doctors of the French Academy who had been invited to the opening ceremonies were first taken to explore the antiquities. Among the luminaries in this group were the painters Jean-Léon Gérôme and Eugène Fromentin and the poet-essayist Théophile Gautier, who left a detailed and sometimes critical account of his experience. He was bemused by the accommodations in Cairo, even though he stayed at Shepheard's Hotel, which would in the late nineteenth and early twentieth centuries become the watering hole of choice for Europeans in Egypt. Like many of his companions, Gautier was mesmerized by the local color, by the snake charmers and the fellah women in their long robes, but he did not take it altogether seriously. Rather than accompany his group to Upper Egypt, Gautier stayed in Cairo so he could mock the pretension of the Gezirah Palace built for Eugénie. "It is of debatable taste," he wrote, "but absurdly sumptuous."[6]

The rest of the party headed south. In theory, the temples of Karnak and the tombs of Luxor had nothing to do with a canal through the isthmus hundreds of miles away. The company, however, understood the power of indirect advertising. The canal benefited from its association with ancient Egypt, however tenuous that may have been. The romance with Egypt throughout Europe meant that people sometimes supported the canal because they were drawn to the pharaohs, and not because they had given serious thought to the cheaper shipment of goods. While aggressively marketing the canal as a more pragmatic route to Asia, the company also presented it as a journey through time. From the outset, Lesseps had shifted nimbly from visionary to pragmatist, speaking in the fulsome, idealistic language of a Saint-Simonian philosopher one moment, and in the accounts-and-ledger patois of a dry-goods merchant the next. No wonder, then, that the company saw an expedition to the ruins of the Nile as a logical prelude to the opening of the canal.

On November 1, Ismail proudly hosted a thousand guests for the opening of the Cairo Opera House. He had hoped to be able to present the premiere of a new opera by Giuseppe Verdi, who was arguably the most prominent composer in Europe. At the urging of Mariette, who had excelled as an archeologist and now aspired to be a librettist, Ismail approached Verdi with an idea for an opera based on Egyptian themes. In spite of the substantial amount of money offered, Verdi was uninterested. His cultural universe did not extend south of Rome, and the idea of premièring a new opera in what was for him the middle of nowhere held no appeal. He was also in the midst of a decades-long love-hate relationship between commerce and his art. "In these days," he wrote to a friend, "art is no longer art but a trade. . . . I feel disgusted and humiliated."[7] He wanted nothing to do with what he took to be a naked publicity gesture. Instead, he agreed to let the khedive mount a production of one of his earlier works, *Rigoletto,* which was dutifully performed during the first weeks of November along with several cantatas and pieces composed especially for the opening. But Mariette was not willing to let go. He persisted and, with Ismail's financial support, finally convinced Verdi to compose an opera based on an apocryphal story of a tragic love triangle in the time of Ramses III. In return for 150,000 francs, Verdi set his qualms aside and went to work. Shortly before Christmas 1871, *Aïda* premiered at the Cairo Opera House. Verdi himself did not attend. He waited instead for the European premiere at La Scala.

On November 16, ships filled the artificial harbor at Port Said, sixty or so vessels crammed into the five-hundred-acre space between the magnificent jetties devised by the brothers Dussaud. Dozens of other ships congregated just outside the enclosed harbor to witness the send-off of the flotilla. Many had just arrived and were bearing excited and skeptical passengers. As Sir Frederick Arrow wrote in an account of his trip, until the actual opening day "the canal was still . . . behind a cloud, and believed by many to be a myth." Some English observers noticed the conspicuous dearth of English flags among the many banners on the shore; others joined the hyperbolic chorus. "The power of man's mind, penetrating and compelling the powers of nature, achieved this," wrote one. "Can it escape the mind of the European who beholds the work that he is standing but a short distance . . . from the spot where a man's arm, animated by the powers of God, smote the hard rock and the waters flowed out?" Port Said looked better than it ever had. Flags were everywhere, and bunting hung from streetlamps and balconies. Ships were lined up, ready to proceed, and at their head was the empress's yacht *L'Aigle* and the khedive's yacht, the *Mahroussah*, which had been the scene of a reception the evening before.

The morning brought a parade of ships, better organized in theory than in practice, but in spite of some jostling and logjams, thousands cheered on the shore and enjoyed the spectacle. Naval vessels from various countries participated in a joint salute to the flags of Turkey, Austria, Egypt, Prussia, and Holland, and smoke and the smell of gunpowder filled the air. The afternoon was given over to a religious ceremony. A landing area had been set aside and decorated with a triumphal arch. On three platforms, arranged around an open square, sat nearly a thousand dignitaries.

Eugénie disembarked and skipped onto the shore, her hand guided by the khedive. The bands, the music, the decorations, and the protocol were not that different from similar festivities in Europe, but the guests were. The thousand who assembled on the platforms formed an international menagerie. The sharif of Mecca, the shaikh of al-Azhar, and the religious scholars of Cairo donned their finest robes and turbans for the occasion, as did their students and retainers. The palace guard of the khedive was in full dress with medals. Kings, princes, ambassadors, and assorted royalty of Europe wore uniforms laden with decorative epaulets and ceremonial swords encrusted with jewels. Queens, princesses, and assorted royal consorts wore the latest finery, with silk and damask fans. Also present was Emir Abd-el Kadir, who had led the Algerian resistance against the French and then became a voice of peace, said some, and a European stooge, said others. That afternoon he occupied a place of honor next to the empress, the khedive, and Lesseps, and people remarked on the contrast between the entrepreneur in his dark suit, and the emir, gray beard barely visible beneath his hood, his robes flowing to the ground.[8]

Watching the ceremony was a large crowd, attired in a cross-section of multicultural fashion, from frock coats and fez caps to Bedouin robes and Prussian helmets. Arab men mixed with Turkish officials, Circassian soldiers, and Armenian merchants, who in turn stood next to Nubian slaves, Greek businessmen from Alexandria, and English seamen. Half-veiled fellah women, with golden ringlets on their headscarves, stood next to French and English ladies equipped with hats and parasols. They all grew silent to hear the sermons delivered by the heads of the respective religious communities. As with everything said in public during those three days, the oratory was bombastic and the imagery was florid. The speakers believed that this was the dawn of a new era, and they treated the occasion with the solemnity they thought it deserved. Some listeners chuckled at the rhetorical excesses; many others were moved.

First was a Muslim benediction given by a stately judge. He spoke softly but proudly of the achievement. The canal may have been conceived by a European company, but much of it had been carved by Egyptian hands, and most of it had been paid for by taxes collected by the Egyptian government. As a salaried official of the state, the shaikh had as much right to claim the canal for Egypt as Lesseps did for France and as the assembled nobles of Europe did for Western civilization. Next, the archbishop of Jerusalem recited a prayer asking God to bless the canal and the safe passage of the flotilla. And then Eugénie's confessor, Marie-Bernard Bauer, gave the last and longest valediction. He spoke of the crescent and the cross, which had fought for centuries and were now being united. He spoke of Asia and Africa, which had "met without touching," and were now closer for being linked by the canal. "Today, two worlds are made one. The splendid Orient and the marvelous Occident salute each other. . . . Today is a great festival for all of humanity." He praised both Islam and Christianity, two faiths which worshiped the same God, no matter how much they had fought in the past. He praised the khedive for his wisdom and courage in supporting the endeavor, and then he turned to Eugénie. "Madame, and these are not idle words, history will say that all of this wondrous work is due to your warm sympathies." Without her support, he told the spectators, the canal might never have been built. Finally, he saluted Ferdinand de Lesseps, "whose name should be placed side by side with Christopher Columbus," because not since that explorer had landed in the New World had any one person so transformed the globe. Father Bauer finished by asking God to "bless this new highway . . . Make of this canal not only a passage to universal prosperity, but make it a royal road of peace and of justice; of the light, and of the eternal truth."

The next morning, the flotilla was ready. But Lesseps had spent a chaotic night. Something had gone terribly awry. The Egyptian pilot ship, one of the most elegant frigates in Ismail's fleet, had steered off course and run aground near Kantara. The ship was blocking the only navigable line of the canal. A barge sent out to tow it failed. Another ship was used as a ram but could not dislodge it. Lesseps conferred with Ismail, who considered the situation and came to the only conclusion: "Blow it up!" Lesseps was relieved, and embraced the khedive. For the moment, they were on the same side. The pressure was intense, and Ismail chose to sacrifice one of his ships for the good of the canal. As it turned out, the frigate was spared, and hundreds of men working until dawn managed to unmoor it. When the flotilla arrived the next day, people had no inkling of the crisis the night before. All they saw was an Egyptian ship of the line saluting them as they glided past.[9]

The procession began with *L'Aigle,* three hundred feet long, with a sixty-foot beam, and not the subtlest of ships. Many years later, a letter-writer to *The Times* claimed that the official story had expunged one embarrassing detail. *L'Aigle* was so lumbering that a small British naval vessel took the opportunity to dart to the head of the line and enter the canal first. The ship was quickly hustled out of the queue and expelled from the flotilla. Eugénie and Lesseps had other reasons for smiling grimly as they passed by the wooden obelisks that had been erected at the canal's entrance. The channel was deep enough, but it had only been dredged across a narrow width. Any ship that deviated from the prescribed course ran the risk of grounding. Lesseps and Eugénie feared that if that happened in broad daylight to the first ship, it would be a "disaster. . . . The national honor of France would be compromised. The future of the canal would be destroyed!" These fears may have been exaggerated, but the captain of *L'Aigle* was nervous; he repeatedly ordered the yacht to stop, and Eugénie paced back and forth from cabin to deck. Eventually, the ship cleared the narrows without incident, albeit behind schedule, and not without considerable muttering up and down the line.

The lead ships stopped at the viewing stands at El-Guisr, north of Ismailia and Lake Timsah. Just behind were the yachts of Ismail, the British ambassador, and the emperor of Austria, after which came other ships carrying everyone from Voisin Bey and his engineers, to the company's board of directors, to the Dutch consul general, Ruyssenaers, to the thousands of guests and tourists in the remaining vessels. Eugénie took a brief tour of the heights of El-Guisr, and was then led back to her yacht, escorted by a carriage pulled by white camels. L'Aigle docked at Ismailia late that evening. The convoy had safely reached the halfway point. It had cleared the narrowest part, and nothing now stood in the way of finishing the trip.

The next day, Ismailia was a carnival. Every public building was given over to the festivities. The khedival palace was closed in preparation for the evening, but Lesseps, Voisin, and Lavalley, and dozens of company officials opened their homes and gardens for visitors. Hundreds of tents and pavilions were erected, some devoted to cooking meals, others to entertainment. As many as ten thousand fellahin and Bedouins camped around the town, to pay their respects both to Ismail and to the French empress, and to enjoy the party. They brought with them some of the atmosphere of Tanta, and Ismailia teemed with tribesmen on camels and men on horses from the deserts of Arabia preparing for a tournament that afternoon.

There were riding exhibitions, tilting with bamboo lances, and a rifle competition. The town, wrote Sir Frederick Arrow, was a whirlwind of "Jews, Turks, infidels and heretics, armed with every variety of weapon. Here a native acrobat, walking the tight-rope, with a baby lashed to each of his ankles—the heads of the poor little couple getting swiftly knocked about. Further on an Armenian trotting out a dancing bear; then comes an Italian with a hurdy-gurdy; then an Arab sword-dance; then a Greek festa. Triumphal arches were in plenty; and as to the illuminations at night, they really made it fairy-land; it was Arabian Nights without the genii."

Cafés were set up to provide an endless supply of demitasse coffee, and pipe boys were kept busy throughout the day and night. Musicians who had traveled for hundreds of miles to perform on the streets competed with glassblowers, flame eaters, snake charmers, and jugglers for attention. Dervishes also arrived, to entertain the Arabs and Egyptians with their dances, and they in turn competed with belly dancers and prostitutes. Koran reciters marked out their turf, and people gathered around fires to listen to mellifluous ballads of Arabic poetry. Everything was paid for by the khedive. No restaurant charged money, and everyone had a place to sleep, though except for the very important persons, it was a challenge for Europeans to find an actual bed to their liking. Food, at least, was not a problem, and there was ample European cuisine, abundant ale, passable wine, good champagne, and even soda water.

On November 18, there was a ball at the palace, which all tried to attend whether they had been invited or not. There was a shortage of carriages in town, and many people simply walked the mile to the chalet in their starched uniforms or tulle dresses. The area was illuminated by Chinese lanterns hung from the palm trees that lined the streets. Arriving at the palace, most attendees found that the number of guests far surpassed the capacity of the rooms. People were crammed into every corner, and the men noticed that there were precious few women. Given the mad crush, a few who showed up were able to take advantage of the rest, and some guests left the ball missing a pocket watch or piece of jewelry. The buffet included sorbets, fruit, and a copious amount of wine, but the line for it was so long that many gave up and found food elsewhere. The khedive and the royals, however, had a separate supper in a formal dining room closed to the public, and their meal included "Fish of the reunion of the two seas," "Beef tongue in the English style, with Nerac aspic, and Suez shrimp salad accompanied by watercress." The nobles finally appeared in the public rooms toward midnight, just after a fireworks display lit up the northern end of the lake and then rained down on the ships moored in the canal. There were formal dances, with Lesseps leading Eugénie, who wore a diamond necklace for the occasion.

The next morning, November 19, was Ferdinand de Lesseps's sixty-fourth birthday. The flotilla departed for the south. Already, under the supervision of Nubar Pasha, the tents were being struck, and a long caravan of camels and donkeys started to make the trip back to Cairo. Some people noted that Ismailia by day was a much less enchanting place than it had been the night before. Some had also had too much to drink. The air was calm, and the currents were gentler than usual. There was confusion, however, about which ships were supposed to go when, and craft of different nationalities competed for position. The plan had been for the entire convoy to reach the Bitter Lakes by late that evening, but, given the chaos and the competition, some ships didn't leave Lake Timsah until the morning of November 20. Even so, with fewer concerns about grounding, the lead ships went more quickly. Though they had to exercise caution in order to pass through the Serapeum, once they reached the Bitter Lakes they spread out and raced one another.

The final test was the narrow channel through the Chalufa ridge, which had until a few weeks before been impassable. From there, it was easy sailing to the port of Suez, which like its sister cities to the north was decorated for the reception. The day was exceptionally beautiful. Writers remarked on the crystal-clear air, which made the Gebel Attaka, at the southern end of the canal, that much more dramatic. The town of Suez, however, didn't impress. "Suez," said one writer, "in spite of its holiday finery, can only be described as a horrible place." Here as well, Ismail had made sure that there were large crowds. An honor guard stood at the point where the canal opened out into the Red Sea, and they fired their guns to salute *L'Aigle* as it approached. The khedive had arrived earlier, and he met the empress's yacht. For the sake of theater, he was rowed by a dozen oarsmen in his state barge from the railway terminus at the water's edge to the side of *L'Aigle* so that he could welcome Eugénie in full view of the crowds. That evening, the town was lit from every house, and the strains of "La Vie parisienne" could be heard in the streets. A final fireworks display greeted the ships still straggling into port. Fifteen years after Lesseps gained the concession from Said Pasha, a decade after the furtive beginning at Port Said, the desert had been parted, and the canal was open.

Everywhere he went during the days, Lesseps was hailed as a hero. Songs and ballads were composed in his honor. But his greatest moment may have come a short while later. There were more festivities in Cairo on November 21 and 22. The khedive sponsored another ball, and then hosted races at the Pyramids. The royals and special guests were taken from Suez to Cairo by train, but others found that the road back was more difficult than the trip going. There weren't enough trains scheduled, and those who tried to make the canal journey north to Ismailia and Port Said found the process chaotic and slow. Surprisingly, though the company had not yet developed an adequate system for guiding ships through the canal, none got stuck except for the Egyptian frigate on the eve of the opening day. By train, by ship, or by camel and on foot, the visitors left the isthmus and headed home, some to Europe, some to their villages along the Nile, and some to the deserts of Sinai and Arabia. As for Lesseps, on November 25 he did something he had done only once before in his life: he became a husband.

The opening had been a success, and the work that had consumed his life for the past fifteen years was complete. Lesseps had become a larger-than-life figure, but he was still just a man who had conceived of the canal only after the death of his wife. For all of its grandeur, the canal had given him something to do, and with its completion, there was the prospect of being alone once again. In a small ceremony in Ismailia, he wed Hélène Autard de Bragard. She was twenty-one years old, the creole daughter of a well-to-do family from the island of Mauritius. Lesseps was three times her age. They would be together for the remaining twenty-five years of his life, and she bore him twelve children. News of the wedding was briefly reported in *Le Figaro*: "M. de Lesseps, after having married the Mediterranean and the Red Sea, then got married himself."

# CHAPTER TWENTY

# THE LEGACY

THE CANAL HAD been successfully navigated by a large flotilla of ships from a number of different countries. News of that success was widely disseminated, and almost no one questioned the technical achievement. Though a fair amount of dredging and widening was still needed, the canal had met or surpassed expectations. "Of this . . . we may feel certain," pronounced one English writer, "the Canal is an established fact. It will disappear no more." Less clear was whether it would draw traffic, and many wondered if the company would survive. Ten francs a ton seemed quite steep if the company wanted the route to be competitive, yet if it charged much less it could not meet its expenses or satisfy shareholder expectations for dividends. The skeptics were right. In the first years of its operation, the Suez Canal was a commercial disaster.

In 1870, its initial year, slightly fewer than five hundred ships made the passage, carrying a total of slightly more than four hundred thousand tons. That was a pittance compared to the five million tons that the company had promised. Great Britain was responsible for three-quarters of what trade there was. The next year was better, with the total rising to 750,000 tons, but this was so far below projections that the company faced insolvency. Shareholders watched as their equity shriveled to two hundred francs a share. Dividends failed to materialize. A pamphlet was published in Paris entitled *The Agony of the Suez Canal. Zero results. Next comes ruin!* To placate angry and restive investors, the company negotiated new loans to meet operating expenses and lowered its rates. Lesseps, accustomed to being greeted at annual shareholder meetings with standing ovations, found himself booed and criticized.

Stung by the rebukes and disappointed by the financial returns,

Lesseps still commanded immense respect. He had received the cross of the Legion of Honor from the empress; he had been decorated and lionized throughout Europe after the opening; and his shares in the Canal Company made him a very rich man. The canal zone continued to prosper, and there was speculation that Ismailia would eventually supplant Cairo as the capital of the khedive's new Egypt. Lesseps settled into the life of a grand patrician, dividing his time between Ismailia, Paris, and a French country estate. He was a celebrity wherever he went, and in spite of his age, he retained the same energy and will that had driven him for the past fifteen years. He socialized with the leading citizens of Paris and the world, and within a few years of the canal's completion, he was looking for another venture.

Events in France enhanced his standing. In the summer of 1870, Napoleon III declared war on Prussia. This conflict, which both sides sought, resulted in one of France's most humiliating defeats. Prussia had gone from a second-tier state to a first-rank power, thanks in no small measure to the tactics of Otto von Bismarck. The Second Empire of Napoleon had gone in the opposite direction. Though his government became less and less autocratic, the economy had stagnated. The army had deteriorated, and political dissent had grown acrimonious. Convinced that he would do justice to the name of Bonaparte, Napoleon led his troops into battle at Sedan that September. Surrounded by the Prussians, he was forced to surrender, and Paris was besieged.

With the emperor a prisoner, the Second Empire collapsed. A new government kept the Prussians from occupying the city and then sued for peace. Soon it was fighting a second front against domestic opponents, and Paris erupted into civil war between the conservative bourgeois government and the radical, utopian leaders of the Paris Commune. Not until the end of 1870 was the Commune defeated, and then only after tens of thousands had died on Haussmann's elegant boulevards. The new government wrote a new constitution, and the Third Republic, with a weak presidency and a powerful Assembly, began. Napoleon III, who had come to power so improbably, lost it almost as dramatically. He died in England in 1873. Eugénie lived for nearly half a century more, until 1920, and Queen Victoria, another longtime widow, became her close confidante. The opening of the Suez Canal proved to be the apogee of her years as empress. The Second Empire became a national question mark for the French, who have never been able to decide whether it was a product of collective folly and corruption, or a necessary period of peace and stasis in an otherwise tumultuous century.

The Third Republic was dominated not by a king and court, but by an assembly of representatives, most of them younger than Lesseps. They were adamant champions of business, industry, and empire, and they tended to view the Suez Canal and the Suez Canal Company as two of France's greatest achievements and Lesseps as a national treasure. After the company decreased its rates, the canal began to attract more freight, and the stock price recovered. Lesseps, who had never flagged, felt vindicated. His advice was sought on numerous projects, and along with his son Charles, he became involved in a new scheme that followed logically from the first but was, in truth, much more complicated and much less feasible: carving a canal across the Isthmus of Panama, and doing for North and South America what he had done for Asia, Africa, and Europe.

At almost the same time that Lesseps announced his goal to construct a Panama Canal, the Suez Canal returned to the center of international politics. After 1869, Ismail went even deeper into debt. In 1873 alone, he borrowed more than thirty million pounds, which was double the cost of building the entire Suez Canal; of that thirty, he received barely twenty million, with the rest due in interest.[1] By some accounts, he spent as much as ten million pounds on bribes to the Ottoman Empire. The rest was used for the ongoing modernization of the country. His public-works projects, though expensive, were altering Egypt, and state revenues increased.

By mid-1870, however, the treasury was nearly a hundred million pounds in debt, and the interest burden alone was crushing. Creditors in France and England became worried. They had lent the money on exorbitant terms, and now wondered if Ismail would be able to meet his obligations. In 1875, European bankers decided that Egypt was on the brink of insolvency. Ismail needed several million pounds in order to pay the next installment on the debt in December, and no lender was willing to extend him further credit. He was in danger of defaulting, and he had one attractive asset: his shares in the Suez Canal Company.

At first, it appeared that he would sell the shares to French bankers. The lead buyer was the Société Générale. Word of the impending transaction reached London. The British prime minister, Benjamin Disraeli, had already investigated the possibility of purchasing the khedive's canal shares that past summer, and when he was informed that the French, who owned half of the canal, were about to buy the rest, he acted quickly. Some of his motivation was political. Though the Third Republic had replaced Napoleon, it embraced his imperialism. French foreign expansion in Africa and Southeast Asia competed directly with British expansion in those same regions. It was a game of global chess, played with real soldiers and actual countries. Having tried to prevent the canal from being built, the British were starting to rely on it, and that raised specters of dependency on French good will. If Parisian bankers allied with Lesseps bought the khedive's shares, Disraeli later explained, "the whole of the Suez Canal would have belonged to the French, and they might have shut it up!" In order to prevent that unlikely occurrence, Disraeli moved swiftly to counter the French offer.

It was a brilliant move, executed deftly, and motivated by an acute reading of the strategic, political, and economic advantages. Had another man been in the same position, he might not have grasped the opportunity, even if he had shared Disraeli's political acumen. Unnoticed at the time, and rarely remarked on thereafter, was the degree to which Disraeli acted not on political principle but on the same passion for the Orient that had animated the Saint-Simonians and Lesseps. Decades earlier, when Muhammad Ali was at the height of his power, Disraeli had made a pilgrimage to Palestine and Egypt, after which he wrote several novels about the Orient. In *Tancred,* which was published in the late 1840s, Disraeli created a protagonist who goes east to find the missing element in his life. The young nobleman Tancred, consumed by "the fever of progress," realizes that science and industrialization are not enough. In the Orient, Tancred discovers the solution: a fusion of East and West, of Asia and Europe, science and religion, that could make both whole. Thirty years after writing *Tancred,* Disraeli found himself in a position to fulfill his own literary prophecy.

He turned to the Cabinet, to the queen, and to Lionel Rothschild. The Conservatives had a majority, and as leader of the party, Disraeli could make an executive decision with a reasonable expectation that the Cabinet would support him. After several hours of debate, it did. Though the Liberals, led by Gladstone, grumbled, they could not stop Disraeli. Not only did he have a parliamentary majority, he also had a singular ability to communicate with Queen Victoria, and she gave him her blessing. She apparently viewed an English purchase of the canal shares as a blow not against the French but against Bismarck, who had made several very public statements questioning whether Britain was still up to the task of being a great empire.

The shares would cost four million pounds. Disraeli was prime minister, but he did not have that sum of money at his immediate disposal. Needing a quick infusion of cash, he approached Lionel Rothschild for a loan. Years before, Ferdinand de Lesseps had asked James Rothschild for help in capitalizing the Canal Company and then turned him down when told how much the aid of the Rothschilds would cost. Disraeli was able to negotiate far better terms and, on behalf of the British government, contracted a loan with Lionel for four million pounds at 5 percent interest and 2.5 percent commission. Four million pounds was slightly more than the French consortium had offered for the khedive's 177,000 shares. The money would be transferred to Ismail by the Rothschilds, and Ismail would then physically deliver his share certificates to the British consulate. The transaction went smoothly, and at the end of November 1875, the British government owned 44 percent of the Suez Canal Company and became the largest single shareholder. Writing to Queen Victoria, Disraeli declared, "You have it, Madam."

The French press reacted as if France had suffered another military defeat. The French government worried that Britain would become more involved in Egypt and start to undermine the French position in North and sub-Saharan Africa. Lesseps, whose last-minute effort to outbid Disraeli had failed, issued a statement chiding the British for having spent years fighting against the company only to turn around and join it. Trying to sound magnanimous, he expressed the hope that Britain and France would now work together for the benefit of world trade. The Germans enjoyed watching the latest setback to French power. But in England, the news was not greeted with universal acclaim. *The Times* worried about the long-term consequences of Disraeli's action. "The possible results of this national investment are so large and indefinite," the paper prophetically editorialized on November 27, 1875, "that it would be vain to speculate upon them. . . . It is plain that we acquire an interest in Egypt and its administration which will compel the constant attention of the Queen's government. We have purchased nearly half the shares of the Suez Canal. . . . To this country will belong the decision on every question, whether scientific, financial or political . . . and as we have the power, so we shall have responsibility before the world. . . . We now have an abiding stake in the security and welfare of Egypt."

Disraeli's actions had almost as much effect on subsequent world history as the building of the canal itself. The four-million-pound infusion temporarily allowed Ismail to satisfy his creditors, but within a year, he was faced with the same problem. With no more shares to sell and no way to pay the interest due, he was forced to accept a joint Anglo-English commission to oversee the management of Egypt's finances. The arrangement, known as the Dual Control, consolidated Egypt's outstanding debts and provided an excuse for a consortium of international bankers to take charge of the treasury. Ismail's government was, in effect, placed in receivership, and he lost the ability to determine his state budget. He had dedicated his life to making Egypt independent of the Ottoman Empire. He had done everything he could to revolutionize Egypt in order to place it on a par with European states. And now he found himself condescended to by bankers who treated him as a child who had spent too freely. Unwilling to accept this humiliation, he plotted to remove the Dual Control and regain his authority.

That backfired. Though Ismail had the power to obstruct the efforts of the Dual Control to collect tax revenue, he was losing legitimacy. Ministers turned against him, including Nubar Pasha, who calculated that an Egypt with Ismail was more likely to fall under complete European control than an Egypt without him. In April 1879, after a series of intrigues that failed to dislodge the Europeans, Ismail was deposed by the new Ottoman sultan, who had been pressured by the English and French to remove the troublesome and increasingly impotent khedive. Replaced by his son, Tawfiq, Ismail went into exile and lived the last fifteen years of his life in a secluded villa near Mount Vesuvius in Italy.

Ismail's ouster did not improve matters. His son was inexperienced and was cast into a situation that few could have managed well. One of the first things the Dual Control mandated was the sale of the khedive's right to 15 percent of the canal's profits. Bought by a French bank, the price was twenty-two million francs, which was less than one million pounds. In later years, that 15 percent would be worth more than a hundred million francs each year. With that sale, the Egyptian government ceased to have a financial stake in the canal that it had financed.

That was only one of the many indignities. The blatant exercise of European power stoked nascent Egyptian nationalism. Under Said and Ismail, the old Turkish ruling class had gradually been forced to share power with native Egyptians. Much like Napoleon in the 1860s, Ismail had opened up the political system and allowed for an assembly with limited powers. In 1880 and 1881, Arab-speaking ministers and army officers made their own bid for prominence in the Egyptian government, and when they were rebuffed by Tawfiq and his coterie of Turkish and Circassian ministers, they rebelled. Claiming to speak for the fellahin, these officers, led by Colonel Ahmad Urabi, assailed Tawfiq as a puppet of the Dual Control. That led to a diplomatic crisis, followed by an armed rebellion. Tawfiq appealed to France and Britain for help, but only Britain sent troops. In the summer of 1882, after an anti-European riot, a British fleet bombarded Alexandria, landed an army, and routed Urabi's forces at Tel al-Kabir.

As part of their offensive, British forces seized the canal. Lesseps denounced the seizure as a violation of the canal's neutrality, but British troops had in fact occupied the entire country. Several years passed before an international convention declared that the Suez Canal should never be closed to ships of any nation by any nation, and in the interim, the British established a protectorate over Egypt. While the khedive ruled in name, along with a prime minister and an assembly, the British government held ultimate authority, and the British consul general, Lord Cromer, was the final arbiter of what the government could and could not do.

In purchasing the canal shares, the British government did, as *The Times* had warned, develop "an abiding stake in the security and welfare of Egypt." But the consequences extended beyond Egypt. By the beginning of the twentieth century, the Suez Canal had become the fulcrum of the British Empire and an excuse for imperial expansion. As the volume of trade increased, the British government began to treat the canal as the most vital, and most vulnerable, point in the whole empire. Suez was seen as the link between England and its overseas colonies in India, Singapore, Hong Kong, and Australia. British officials were so concerned about the possibility of Suez's falling into hostile hands that they justified expansion into Afghanistan, along the coast of East Africa, and into Iran and the Middle East. The logic, however tenuous, was that if adversaries such as Russia, Germany, or France controlled any of these regions, they would be able to threaten the Suez

Canal, and if they seized the canal, then the entire British Empire could be severed and dismembered. Operating under that assumption, the British occupied Egypt after the outbreak of World War I and ruled the country directly.

Well before that happened, the final major figure in the canal's construction also came to a less than happy end. Ferdinand de Lesseps and his son Charles founded a company in 1880 to dig a canal across Panama. Lesseps figured that what had worked for Suez would work for Central America, and the new company mimicked the old. It was called the Compagnie Universelle du Canal Interocéanique de Panama. After a few false starts, shares were offered to the public, and the company began publishing a regular paper called *The Bulletin of the Interoceanic Canal*, modeled after *The Isthmus of Suez: The Journal of the Union of the Two Seas*. "I maintain that Panama will be easier to make, easier to complete, and easier to keep up than Suez," Lesseps declared, and once again he drew on the expertise of the engineers of the Polytechnic. He assured investors that the work would be completed by 1892.[2]

By 1889, the company was in shambles. It had spent nearly one billion francs, and had almost nothing to show for it. In truth, Panama was a more complicated and dangerous enterprise. The jungles made the proposed route a zone of disease, rain, and heat. Thousands died. The differential height between the Atlantic and the Pacific meant that not only was more excavation required, but the canal would need locks—a reality that Lesseps only conceded after several years of futile effort. Locks presented technical difficulties that neither Lesseps nor any of the engineers had faced, and cost estimates soared. In 1889, another exposition was held in Paris, and France witnessed two things it had thought impossible: a huge erector set known as the Eiffel Tower, and the downfall of Ferdinand de Lesseps.

The Panama Canal company went bankrupt and was liquidated. Though the individual losses to the half-million shareholders were manageable, the scandal and the disgrace were not. In 1892, Lesseps, his son, and several others, including the recently lauded Gustave Eiffel, were investigated by the Assembly and then charged with fraud and conspiracy. Charles de Lesseps, Eiffel, and the other company directors were arrested and placed in solitary confinement. Ferdinand was kept from jail because, at the age of eighty-seven, he was too ill to be moved from his bed at his country estate. The prosecuting attorney charged the elder Lesseps with having deceived France with "beautiful illusions," and both father and son were sentenced to five years in prison. The court did not enforce the sentence against Ferdinand, and he remained in his bed, only vaguely aware of his utter humiliation. Though the board of the Suez Canal Company issued a statement defending Lesseps and his legacy, he remained mute, and did not leave his room until his death on December 7, 1894.

The Panama Canal was eventually built by an American company at far greater cost. The Suez Canal, meanwhile, assumed ever-greater prominence in world affairs. During World War I, the British stationed more than a hundred thousand troops in the canal zone. To protect the canal, and undermine Turkey, the British sponsored an Arab revolt, and a brilliant, eccentric young officer named T. E. Lawrence crossed the canal into the Arabian Peninsula to help the sharif of Mecca. After World War I, the British governed Iraq, Jordan, and Palestine, and increased their military presence in the canal zone.

Even when Egypt was declared independent in 1922, with a newly crowned King Fuad I, Britain retained the right to defend the Suez Canal and garrisoned troops in the isthmus for that purpose. A substantial percentage of world trade was passing through the canal, and the company's profits soared. With the rise of Egyptian nationalism in the 1920s and 1930s, the British granted the Egyptians more say over how the canal was managed, and the company agreed to pay an annual rent.

During World War II, the British once again clamped down on dissent and occupied the country. Egypt became a significant theater of the war. First the Italians and then the Germans, commanded by Erwin Rommel, tried to evict Britain from Egypt and capture the strategically important canal. Even if the Germans failed to occupy the isthmus, they could do almost as much damage by gaining air supremacy in Egypt and then using their planes to bring shipping in the eastern Mediterranean to a halt. Without the canal, Britain would be cut off from India and Asia. Until the summer of 1942, Rommel's Afrika Korps moved steadily eastward from Libya. Only after the tank battles of El Alamein were the Germans halted, and only then did the threat to the canal recede.

On July 23, 1952, King Farouk, the last descendant of Muhammad Ali to govern Egypt, was overthrown by an officers' coup. Three days later, he was sent, like his grandfather Ismail, into exile in Italy, and he left on the same yacht, the *Mahroussah,* which had been moored in

Port Said when the Suez Canal was inaugurated and which now carried Farouk to a life of dissolute luxury.

Within a few years, Colonel Gamal Abdel Nasser emerged as the leader of the officers who had organized the coup. He was the dominant figure in the Arab world for the next fifteen years, espousing his own particular mix of nationalism and socialism. He called for Arab unity, and refused to take sides in the emerging Cold War between the United States and the Soviet Union. Having come to power on a platform of nationalism and independence from the West, he looked at the Suez Canal as an egregious reminder of Egypt's long decades under British control.

The canal concession was scheduled to expire on November 17, 1968, ninety-nine years after the opening date. But the company still operated as a quasi-autonomous state on Egyptian soil, and still was Egypt's largest source of foreign revenue. The canal accounted for a sizable share of world oil shipments from the Middle East. By the mid-1950s, 122 million tons of cargo passed through the canal annually, and more than seventy-five million of that was oil. Compared with the days when the company hoped to get five million tons of traffic, the canal had come far.

In 1956, the World Bank in conjunction with the U.S. government of Dwight Eisenhower turned down Nasser's request for a loan to build a high dam across the Nile at Aswan. Nasser responded with a dramatic speech on July 26, 1956, exactly four years after Farouk had been sent into exile. Though he spoke in front of a large crowd in Alexandria, Nasser's words were broadcast throughout the Arab world on radio. His long, colloquial speech presented the past century of Egyptian history as a series of struggles against the West that had culminated in the glorious victory of the 1952 coup. Now, Nasser declared, only one thing remained in Egypt's way. Recently, he reminded his audience, he had participated in negotiations for a loan, and the president of the World Bank, Eugene Black, had been in the room. Nasser continued: "I started to look at Mr. Black, who was sitting in a chair, and I saw him in my imagination as Ferdinand de Lesseps."[3] That seemingly innocuous reference was a signal to Egyptian forces, who streamed into the canal zone and surprised the startled British troops. Nasser concluded his speech with the declaration that, as of that moment, Egypt had nationalized the Suez Canal Company. For the first time, the canal belonged solely to Egypt.

It was a short-lived victory. In October 1956, Britain, France, and Israel responded by attacking Egypt. Large portions of Port Said were destroyed. British forces retook parts of the canal, and in retaliation, an angry crowd of Egyptians blew up the thirty-five-foot-high statue of Ferdinand de Lesseps that stood at the canal's entrance. The United States refused to countenance what the British, French, and Israelis had done, and the canal was then returned to Egypt. For England especially, it was a signal defeat. The British Empire, it was said, came to an end with the Suez crisis, and Prime Minister Anthony Eden resigned in disgrace. The crisis also altered the fortunes of the canal itself. No longer the center of an empire, the Suez Canal was now one point among many in the global Cold War, and hardly the most important. That didn't matter to the Egyptians, however. For the next ten years, Egypt operated the waterway and negotiated a compensation agreement with the canal-less Suez Canal Company. Though its international profile dwindled, the canal continued to provide Egypt with dearly needed foreign currency.

In 1967, Israel went to war with Egypt again, and this time, all three cities along the isthmus were devastated by Israel's air force and artillery. Israeli troops and tanks crossed the canal, and only retreated to the eastern shore after the armistice. Nasser ordered dozens of ships to be scuttled to make sure that no one could use the passage, and the canal remained closed until 1975. Soon after, Israel and Egypt signed an accord at Camp David, and Israel returned the Sinai Peninsula and the eastern shore of the canal to Egypt in return for a peace treaty recognizing Israel's existence.

The canal reopened, and it was deepened and widened to accommodate large oil tankers. But after 1975, history began to reverse itself. Since its inauguration in 1869, the Suez Canal had enjoyed a steady expansion of its business. Now it started to contract. By the 1980s, pipelines had taken away a large portion of the canal's oil business, and a new generation of tankers were too big for the waterway. World shipping began to bypass the canal, and in order to retain traffic, the Egyptian government was forced to cut rates. This was a losing game.

The total number of ships passing through the Suez Canal peaked in the early 1980s at twenty-two thousand annually, and had fallen below fifteen thousand in 2002. At the turn of the millennium, the canal still generates nearly two billion dollars a year in revenue for Egypt, but that figure has ceased to grow even as the country's population balloons.

The cities along the canal zone have become more populous, along with the rest of Egypt, and the government has put hundreds of thousands of people to work for the Suez Canal Authority. But there is an inverse relationship between the number of people employed and the amount of work to be done. The canal cities have started to resemble mini-Cairos, overpopulated and filled with young men and women who have just enough work, food, and subsidized housing, and little possibility of anything else. There is chronic talk of multibillion-dollar projects to dredge the canal, widen it, and make it more attractive to international shipping. But with seventy million people, many of them disillusioned by a repressive government and decades of an unresolved Arab-Israeli conflict, and a few of them drawn to the eschatological appeal of radical Islam, Egypt has enough trouble maintaining its precarious status quo. The canal still functions. It is still a testament to nineteenth-century will and ingenuity. But its legacy for Egypt is a different, and sadder, story.

# OZYMANDIAS

I N LATE DECEMBER 1817, two young English poets held a contest. They would each write a sonnet based on a reference made by the ancient historian Diodorus Siculus. The subject was a statue of Ramses II, who was widely thought to be the pharaoh of the book of Exodus who tried to keep Moses from leading the Israelites out of Egypt. Then, under pen names, they would both submit their final poems to *The Examiner,* a prominent London journal, and see who received the greatest acclaim. One of the two was Horace Smith; the other was Percy Bysshe Shelley.

Neither man had been to Egypt, but they were excited by what they had heard. Pharaonic Egypt fit their sensibilities perfectly, and the image of a colossal statue in the sand was irresistible. According to Diodorus, a thousand-ton statue of Ramses had been erected near the Theban necropolis, on the west bank of the Nile near Luxor. There was an inscription on the statue's base, which declared that no one would be able to surpass the achievements of Ramses the Great. Yet, centuries later, as Shelley and Smith knew, that statue (or at least what explorers at the time believed was that statue) was in ruins. It lay prostrate in the desert, its face smoothed by centuries of wind and sand, the inscription almost erased by time. Smith wrote a short poem with a long title, called "On a Stupendous Leg of Granite, Discovered Standing by Itself in the Deserts of Egypt, with the Inscription Inserted Below." His friend Shelley offered a pithier title, the Greek word for "Ramses," or "Ozymandias."

The conceit of the sonnet was that Shelley had met Diodorus and had been told of the statue. For the passionate, dreamy Shelley, Ozymandias was a metaphor for fate. For him, as for so many of the

Romantics, all beauty is transitory, flaring brightly before it is consumed. All that is left is the pain of loss and the poignant memory of a glorious moment. Human beings, weak as they are, always forget that truth. They succumb to the illusion that they can alter history and be immortal, and that sets them up, age over age, and era after era, for the inevitable fall.

> *I met a traveller from an antique land*
> *Who said: Two vast and trunkless legs of stone*
> *Stand in the desert. Near them, on the sand,*
> *Half sunk, a shattered visage lies, whose frown,*
> *And wrinkled lip, and sneer of cold command,*
> *Tell that its sculptor well those passions read*
> *Which yet survive, stamped on these lifeless things,*
> *The hand that mocked them, and the heart that fed:*
> *And on the pedestal these words appear:*
> *"My name is Ozymandias, king of kings:*
> *Look on my works, ye Mighty, and despair!"*
> *Nothing beside remains. Round the decay*
> *Of that colossal wreck, boundless and bare*
> *The lone and level sands stretch far away.*

In November 1854, Ferdinand de Lesseps and Said Pasha imagined a canal that would part the desert. They believed it would be a work for the ages. Like the Saint-Simonians, they saw the canal as the fulfillment of human potential, as a bridge between worlds, and as a path to progress. Lesseps had promised Said that "the names of the Egyptian sovereigns who erected the Pyramids, those useless monuments of human pride, will be ignored. The name of the Prince who will have opened the grand canal through Suez will be blessed century after century for posterity." The canal would increase the wealth of Egypt and strengthen the bonds between civilizations. That vision nurtured Lesseps, and it offered succor to all who dedicated themselves to turning the idea of the Suez Canal into a reality.

As Shelley keenly grasped, human history is littered with statues of Ozymandias. People become convinced that they have found the key to greatness. They create works of art; they conquer; they build invincible armies; and they tame the earth. They construct buildings meant to last forever, and philosophies designed to answer the eternal questions. Ferdinand de Lesseps, Enfantin, Louis-Napoleon, Said, and Ismail constructed the Suez Canal with that same surety. They expected future generations to "look upon their works" in awe at their achievement.

Though he died decades before the canal was built, Shelley could have predicted the outcome. He would have smiled at the hubris, and nodded in empathy, and he would have watched wistfully as Napoleon III went into exile, as Ismail sailed off on his yacht, as Egypt was occupied, Lesseps disgraced, Great Britain triumphant and then humbled, and then, finally, the canal itself, once the apex of the world, receding from history.

Visionaries created the canal, but others actually built it. The fellahin who were brought to the isthmus by the corvée probably did not share the sense that they were involved in a great undertaking. They had a collective memory of forced labor in the past, and they could not see how their lives would be much the better because of a ship canal. Rulers had always had their passions and their follies; the Nile, after all, was littered with such rulers' remains. After the corvée, the immigrant laborers who worked on the canal in the 1860s came because there were jobs to be had. If the ultimate result improved the world, all the better, but that wasn't what drew them there. The merchants and shippers who took advantage of the canal were not prone to think in metaphysical terms. If the canal shortened the trip and lessened the cost, they would use it. And as for the marriage between East and West, most of them, it is safe to say, cared little if it ended in divorce.

The Suez Canal was the greatest feat of organization and engineering of its day, and it served, for a brief moment, as a symbol of all that was right in the world. It was created by dreams and by meticulous organization, by brilliant engineers and by workers looking for their next meal. And then, once the fireworks had faded, the canal began to fade as well. Traveling through Suez today, it is tempting to despair. Barbed wire, overpopulation, rusting ships, and dwindling business stand as rebukes to the vision of Lesseps.

And yet Shelley himself never gazed at that colossal statue lying on its side near the tombs of the pharaohs. He never saw that, even in decay, there is something stirring about standing there and staring at the ruin. It is a wreck, true, but it is awesome. Its ability to survive across thousands of years inspires a sense of wonder, and it is made greater by its ruin. Had it been seen in its own time, it would have been a large statue of a king, impressive but not unique. Thousands of years later, it has become a record of human history, its impact magnified by the humility it demands.

The canal sits now, wider than it was, deeper than it was, but still flowing along the same course, from Port Said, past the marshes of Lake Manzala, between the cliffs of El-Guisr, into Lake Timsah and past Ismailia, narrowing through the Serapeum before opening out into the Bitter Lakes, and then funneling through the Chalufa ridge before its last, gentle passage into the Gulf of Suez and the Red Sea. It is polluted, and the landscape is scarred from successive wars. Sitting at the point where the canal ends and the Red Sea begins, watching dilapidated freighters glide past, it is hard not to focus on decline, but that is too easy. The Suez Canal was the inscription of an idea on the face of the earth. As a vision, it was beautiful and inspiring; as a reality, it has sometimes been a blessing, and usually not. In its prime, it offered, at best, power and wealth. In its decay, it is Ozymandias.

# Notes

A *note on the notes:* Rather than clog the manuscript with hundreds of reference numbers, the notes are usually a compilation of several citations. For the most part, there is a clear indication of where quotations have come from, but in many cases, for the sake of simplicity, I have grouped together a series of primary and secondary references, especially for biographical sketches and for the final chapters.

## CHAPTER ONE: THE TWILIGHT

1. Account is taken from the journal *L'Isthme de Suez,* May 15, 1859, in Suez Canal Company Archives (Archives d'entreprise, la Compagnie universelle du canal maritime de Suez), Centre des Archives du Monde du Travail, Roubaix, France 1995060-1522. [Hereafter, all references to the Canal Company archives will be abbreviated as CAMT. In addition, unless otherwise stated, all reference numbers for the papers of the company begin with 1995060, and only the rest of the numbers will be given.] All translations from French sources are the author's unless otherwise noted.

## CHAPTER TWO: THE FRENCH FALL IN LOVE

1. For a general account of Louis XIV, see Pierre Goubert, *Louis XIV and Twenty-Million Frenchmen,* trans. Ann Carter (New York: Vintage, 1966). Leibniz's memo to the king was originally written in Latin; see Leibniz, "Consilum Aegyptiacum," trans. into French by A. Vallet de Viriville, in appendix, Ahmed Yousseff, *La Fascination de L'Égypte* (Paris: L'Harmattan, 1998). For the second Leibniz quotation, see J. M. Thompson, *Napoleon Bonaparte* (Oxford: Blackwell, 1952), p. 107.

2. For Volney, see Jean-Marie Carré, *Voyageurs et écrivains français en Égypte* (Cairo: Institut Français d'Archéologie du Caire, 1956), pp. 91–116; Albert Hourani, *A Vision of History* (Beirut, 1961); Volney quotation from Robert Solé, *L'Égypte, passion française* (Paris: Éditions du Seuil, 1997), pp. 36–37. For Napoleon and Talleyrand, see Robert Asprey, *The Rise of Napoleon Bonaparte* (New York: Basic Books, 2000), pp. 50ff; Alan Schom, *Napoleon Bonaparte* (New York: HarperCollins, 1997), pp. 93ff;

Henry Laurens, *L'Expédition d'Égypte* (Paris: Armand Colin, 1989); Alain Silvers, "Bonaparte and Talleyrand: The Origins of the French Expedition to Egypt in 1798," *American Journal of Arabic Studies*, vol. 3 (1975).

3. Napoleon to Josephine quoted in Asprey, p. 253; Napoleon to his men, quoted in Thompson, *Napoleon*, p. 109.

4. Abd al-Rahman al-Jabarti, *Al-Jabarti's Chronicle of the First Seven Months of the French Occupation of Egypt*, ed. and trans. S. Moreh (Leiden: E. J. Brill, 1975), pp. 43–57; see also André Raymond, *Cairo*, trans. Willard Wood (Cambridge, Mass.: Harvard University Press, 2000); Darrell Dykstra, "The French Occupation of Egypt," in *The Cambridge Modern History of Egypt*, vol. 2, ed. M. W. Daly (New York: Cambridge University Press, 1998), pp. 113–38.

5. Figures and description of the expedition as well as Monge quotation from Schom, *Bonaparte*, pp. 93–96. See also Carré, *Voyageurs*, pp. 143–65; Lesley and Roy Adkins, *The Keys of Egypt: The Race to Read the Hieroglyphs* (New York: Harper-Collins, 2000), pp. 23–37.

6. Abd al-Rahman al-Jabarti, *Journal d'un notable du Cairo durant l'expédition française, 1798–1801,* trans. into French and annotated by Joseph Cuoq (Paris: Albin Michel, 1979), pp. 90–95; for al-Azhar pillaging, see Jabarti, *Al-Jabarti's Chronicle*, pp. 100–102.

7. Solé, *L'Égypte*, pp. 54–69; Lord Kinross, *Between Two Seas: The Creation of the Suez Canal* (New York: William Morrow, 1969), pp. 14–19; Napoleon to Le Père, quoted in French in J. E. Nourse, *The Maritime Canal of Suez* (Washington, D.C.: Philp & Solomons, 1869), pp. 10–12.

8. Yousseff, *Fascination,* passim; Solé, *L'Égypte*, pp. 70–80; Michel Dewachter and Alain Fouchard, eds., *L'Égyptologie et les Champollion* (Grenoble: Presses Universitaires de Grenoble, 1994); Lesley and Roy Adkins, *The Keys of Egypt* (New York: HarperCollins, 2000).

CHAPTER THREE: INDUSTRY AND THE SAINT-SIMONIANS

1. Frank Manuel, *The New World of Henri Saint-Simon* (Cambridge, Mass.: Harvard University Press, 1956), pp. 20–22, 79, and passim; Theodore Zeldin, *France 1848–1945: Politics and Anger* (London: Oxford University Press, 1973), pp. 66–74, though Zeldin's account has been widely criticized as unnecessarily scornful; Francis Démier, *La France du XIX siècle* (Paris: Éditions du Seuil, 2000), pp. 110–11; D. G. Charlton, *Secular Religions in France 1815–1870* (London: Oxford University Press, 1963), pp. 38–66.

2. Quoted in Georg Iggers, trans., *The Doctrine of Saint-Simon: An Exposition, First Year, 1828–1829* (Boston: Beacon Press, 1958), p. 203; see also Robert Carlisle, *The Proffered Crown: Saint-Simonianism and the Doctrine of Hope* (Baltimore: Johns Hopkins University Press, 1987), pp. 45–48.

3. Ghislain de Diesbach, *Ferdinand de Lesseps* (Paris: Perrin, 1998), pp. 51–52; Robert Solé, *L'Égypte, passion française*, pp. 126–35.

4. Carlisle, *Proffered Crown*, pp. 180–82; quotation on the Golden Age from

Charlton, *Secular Religions*, p. 69; Hippolyte Castille, *Le Père Enfantin* (Paris, 1859), pp. 4–20; quotation about Jesus and Moses from Jean-Noel Ferrié, "Du saint-simonisme à l'islam," in Magali Morsy, ed., *Les Saint-Simoniens et l'orient* (Aix en Provence: Édisud, 1989), p. 161.

5. Enfantin to Sainte-Pélagie, Jan. 25, 1833, quoted in Philippe Régnier, "Le Mythe oriental des saint-simoniens," in Morsy, ed., *Saint-Simoniens*, p. 29.

6. Poem reproduced in Régnier, "Mythe," p. 40. For more on Enfantin and the trip to Egypt, see Jehan d'Ivray, *L'Aventure saint-simonienne et les femmes* (Paris: Libraire Félix Alcan, 1928).

7. Roger Owen, *The Middle East in the World Economy, 1800–1914* (New York: I. B. Tauris, 1993), pp. 86–88; E. M. Forster, *Alexandria* (New York: Oxford University Press, 1986).

8. Quoted in Annie Rey-Goldzeiguer, "Le Projet industriel de Paulin Talabot," in Morsy, ed., *Saint-Simoniens*, pp. 98–99.

9. Henry Dodwell, *The Founder of Modern Egypt* (Cambridge, Eng.: Cambridge University Press, 1931); Afaf Lutfi al-Sayyid Marsot, *Egypt in the Reign of Muhammad Ali* (Cambridge: Cambridge University Press, 1984); Khaled Fahmy, *All the Pasha's Men: Mehmed Ali, His Army, and the Making of Modern Egypt* (Cambridge, Eng.: Cambridge University Press, 1997). There are different accounts of the slaughter of the Mamelukes, and some versions have them being killed en masse in a narrow alley leading up to the Citadel.

10. See Ghislaine Alleaume, "Linant de Bellefonds et le saint-simonisme en Égypte," in Morsy, ed., *Saint-Simoniens*, pp. 110–20; Marcel Kurz and Pascale Linant de Bellefonds, "Linant de Bellefonds: Travels in Egypt, Sudan, and Saudi Arabia," in Paul Starkey and Janet Starkey, eds., *Travellers in Egypt* (London: I. B. Tauris, 1998), pp. 61–70.

11. Morsy, ed., *Saint-Simoniens*, passim; Solé, *L'Égypte*, pp. 130–35; Diesbach, *Lesseps*, pp. 55–57; Jean-Marie Carré, *Voyageurs et écrivains français en Égypte*, pp. 263–70.

CHAPTER FOUR: A MAN, A PLAN, A CANAL

1. This story is told in Pierre Crabites, *Ismail: The Maligned Khedive* (London: George Routledge & Sons, 1933), p. 4.

2. George Edgar-Bonnet, *Ferdinand de Lesseps: Le Diplomate, le créateur de Suez* (Paris: Libraire Plon, 1951), pp. 1–10. Also, for his seeming invincible optimism, see Edwin de Leon, "Ferdinand de Lesseps and the Suez Canal," *Putnam's Magazine*, June 1869.

3. Charles Beatty, *De Lesseps of Suez: The Man and His Times* (New York: Harper and Brothers, 1956), pp. 28–29; Duff Cooper, *Talleyrand* (New York: Grove Atlantic, 2001).

4. Henry Kissinger, *Diplomacy* (New York: Simon & Schuster, 1994); M. S. Anderson, *The Eastern Question* (London: Macmillan, 1986); L. Carl Brown, *International Politics and the Middle East* (Princeton: Princeton University Press, 1984).

5. Afaf Lutfi al-Sayyid Marsot, *Egypt in the Reign of Muhammad Ali*, pp. 90–92.

6. The first quotation is Palmerston to Lord Granville, May 27, 1839, in Jaspar Ridley, *Lord Palmerston* (London: Constable, 1970), p. 222. The second quotation is from Evelyn Ashley, *The Life and Correspondences of Viscount Palmerston* (London: Richard Bentley & Sons, 1879), p. 381. See also Vernon Puryear, *France and the Levant from the Bourbon Restoration to the Peace of Kutiah* (Berkeley: University of California Press, 1941).

7. The quotation is from George Macauley Trevelyan, *British History in the Nineteenth Century and After* (New York: David McKay, 1937), p. 292. See also E. J. Hobsbawm, *The Age of Revolution 1789–1848* (New York: New American Library, 1962); Francis Démier, *La France du XIX siècle*, pp. 214–47.

CHAPTER FIVE: EGYPT AND ROME

1. Quoted in Albert Hourani, *Arabic Thought in the Liberal Age, 1798–1939* (Cambridge, Eng.: Cambridge University Press, 1983), p. 52. Other sketches of Muhammad Ali can be found in Afaf Lutfi al-Sayyid Marsot, *Egypt in the Reign of Muhammad Ali*; Henry Dodwell, *The Founder of Modern Egypt*; P. J. Vatikiotis, *The History of Egypt*, 3rd ed. (Baltimore: Johns Hopkins University Press, 1985); Khaled Fahmy, *All the Pasha's Men*; Fahmy, "The Era of Muhammad Ali Pasha," in *The Cambridge History of Egypt*, vol. 2, pp. 139–80; Jack Crabbs, *The Writing of History in Nineteenth-Century Egypt: A Study in National Transformation* (Cairo: American University in Cairo Press, 1984), pp. 62–68; and a contemporary account written in the 1830s, Edward William Lane, *Description of Egypt* (Cairo: American University in Cairo Press, 2000), pp. 104–59.

2. Edward Said, *Orientalism* (New York: Pantheon, 1977); Edward Said, *Culture and Imperialism* (New York: Alfred A. Knopf, 1993); Bernard Lewis, *The Muslim Discovery of Europe* (New York: W. W. Norton, 1982); Bernard Lewis, *The Middle East and the West* (New York: Harper & Row, 1964).

3. F. Robert Hunter, *Egypt Under the Khedives, 1805–1879* (Pittsburgh: University of Pittsburgh Press, 1984), pp. 10–35; Roger Owen, *The Middle East in the World Economy, 1800–1914*, pp. 65–91; Edward William Lane, *An Account of the Manners and Customs of the Modern Egyptians* (London, 1836); Groupe de Recherches et d'Études sur le Proche-Orient, *L'Égypte au XIX siècle* (Paris: Éditions du centre national de la recherche scientifique, 1982), passim.

4. Quoted from al-Tahtawi's chronicle of his time in Paris, in Bernard Lewis, ed., *A Middle East Mosaic* (New York: Random House, 2000), pp. 46–47; see also Hourani, *Arabic Thought*, pp. 68–73.

5. Quoted in George Edgar-Bonnet, *Ferdinand de Lesseps*, p. 156. Other sources on Waghorn include papers in FO 97/411, Public Records Office, Kew, London [hereafter PRO]; John Marlowe, *World Ditch: The Making of the Suez Canal* (New York: Macmillan, 1964), pp. 28ff; Dodwell, *Founder*, pp. 30–40.

6. Marlowe, *World Ditch*, p. 45.

7. Letters between Enfantin and Linant, between Enfantin and Negrelli, and

from Enfantin to Emperor Napoleon III, Sept. 10, 1855, all in Fonds Enfantin, Group 7836, Arsenal Library, Paris [hereafter Fonds Enfantin]; report from Consul Murray to Lord Palmerston, May 27, 1847, in FO 97/411 PRO; Annie Rey-Goldzeiguer, "Le Projet industriel de Paulin Talabot," pp. 95–103; Edgar-Bonnet, *Lesseps*, pp. 170–87; Marlowe, *World Ditch*, pp. 44–51; Lord Kinross, *Between Two Seas*, pp. 47–55; Ghislain de Diesbach, *Ferdinand de Lesseps*, pp. 114–19.

8. Marsot, *Egypt*, pp. 87–90; Vatikiotis, *History of Egypt*, pp. 70–73; Emine Foat Tugay, *Three Centuries: Family Chronicles of Turkey and Egypt* (London: Oxford University Press, 1963), pp. 98–100. Also see Ehud Toledano, *State and Society in Mid-Nineteenth-Century Egypt* (New York: Cambridge University Press, 1990) for a discussion of how Abbas got much of his bad reputation after he died.

9. Ferdinand de Lesseps himself wrote an extensive, albeit temperate, account of his mission to Rome: *Recollections of Forty Years*, trans. C. B. Pitman (New York: D. Appleton, 1888), pp. 3–118. See also Charles Beatty, *De Lesseps of Suez*, pp. 62–70.

10. Edgar-Bonnet, *Lesseps*, pp. 80–95; Beatty, *De Lesseps*, pp. 70–72; Diesbach, *Lesseps*, pp. 116–22.

11. Lesseps, *Recollections*, p. 118.

12. Quoted in Diesbach, *Lesseps*, p. 122.

13. Lesseps, *Recollections*, pp. 152–55.

CHAPTER SIX: A JOURNEY IN THE DESERT

1. Quoted in Jack Crabbs, *The Writing of History in Nineteenth-Century Egypt*, p. 92. On Said, see "Said Pacha of Egypt," *Harper's New Monthly Magazine*, vol. 39 (1869), pp. 41–52; F. Robert Hunter, *Egypt Under the Khedives, 1805–1879*, passim; Edward Dicey, *The Story of the Khedivate* (London: Rivingtons, 1902), passim; *Cambridge Modern History of Egypt*, vol. 2, passim; Emine Foat Tugay, *Three Centuries*, pp. 100–102.

2. David Landes, *Bankers and Pashas* (New York: Harper & Row, 1958), pp. 85–101.

3. Ibid., pp. 90–100. Story of the iron paving from Nubar Pacha, *Mémoires de Nubar Pacha* (Beirut: Librairie de Liban, 1983), pp. 153–54.

4. Alexander Scholch, "The Formation of a Peripheral State: Egypt 1854–1882," and Peter Gran, "Late 18ᵗʰ–Early 19ᵗʰ Century Egypt: Merchant Capitalism," both in Groupe de Recherches et d'Études sur le Proche-Orient, *L'Égypte au XIX siècle*; Byron Cannon, *Politics of Law and the Courts in Nineteenth-Century Egypt* (Salt Lake City: University of Utah Press, 1988).

5. Ferdinand de Lesseps, *Lettres, Journal et documents pour servir à l'histoire du Canal de Suez 1854–1856* (Paris: Didier, 1875), pp. 1–5. These same letters are translated in Ferdinand de Lesseps, *Recollections of Forty Years*. Lesseps's letter to Abbas of April 18, 1853, in CAMT, 1164.

6. Ghislain de Diesbach quotes from a 1931 history of the Saint-Simonians that alleges that Lesseps did get blueprints from Arlès-Dufour, and that is, as the next

chapter shows, precisely what the Saint-Simonians accused Lesseps of in 1855. Lesseps's official biographer, George Edgar-Bonnet, makes the same claim, based on a letter between Lesseps and Arlès-Dufour written in 1855. It is credible given subsequent events, but difficult to establish beyond a reasonable doubt. See Ghislain de Diesbach, *Ferdinand de Lesseps,* p. 124.

7. This and all subsequent descriptions of Lesseps and Said from Nov. 1854 are taken from Lesseps, *Lettres 1854–1856* and *Recollections.*

8. Lesseps, diary, Nov. 15, 1854, in *Lettres 1854–1856,* p. 17.

9. Instead, I have used the French original: Lesseps to Muhammad Said, Nov. 1854, in 7836 Fonds Enfantin. Inadequate translation in Lesseps, *Recollections,* pp. 170–75.

CHAPTER SEVEN: WHOSE CANAL?

1. Ferdinand de *Recollections of Forty Years,* pp. 178–97; English copy of Lesseps's letter to Bruce, Nov. 27, 1854, in FO 78/1156 PRO; French copy in 7836 Fonds Enfantin.

2. Text of the concession, Nov. 30, 1854, in Percy Hetherington Fitzgerald, *The Great Canal at Suez* (London, 1876), pp. 293–96. For tensions between Mougel and Linant, see Nathalie Montel, *Le Chantier du Canal de Suez* (Paris: Éditions en Forma, 1998), pp. 29–33.

3. Though some have questioned how revolutionary the Crédit Mobilier was, it was nonetheless an innovative banking system. See Niall Ferguson, *The House of Rothschild: Money's Prophets 1798–1848* (New York: Viking, 1998); Niall Ferguson, *The House of Rothschild: The World's Bankers 1849–1999* (New York: Viking, 1999); David Landes, *Bankers and Pashas,* p. 50ff; Pierre Miquel, *Le Second Empire* (Paris: Perrin, 1998), pp. 116–44; Francis Démier, *La France du XIX siècle,* pp. 258–62.

4. Lucien Jeanmichel, *Arlès-Dufour: Un Saint-Simonien à Lyon* (Lyon: Éditions Lyonnaises d'Art et d'Histoire, 1993).

5. Lesseps to Arlès-Dufour, Nov. 30, 1854, in 7837 Fonds Enfantin; list of founding subscribers, in 7836 Fonds Enfantin.

6. Lesseps to Arlès-Dufour, Dec. 14, 1854, in 7837 Fonds Enfantin; an abridged copy of the letter is also in Ferdinand de Lesseps, *Lettres (1854–1856),* p. 57.

7. Lesseps, *Recollections,* pp. 206–17; see also Bruce Feiler, *Walking the Bible* (New York: William Morrow, 2001).

8. Lesseps's instructions to Linant and Mougel, Jan. 16, 1855, in 7836 Fonds Enfantin.

9. Enfantin to Negrelli, Jan. 18, 1855, in 7837 Fonds Enfantin.

10. Lesseps to Arlès-Dufour, Jan. 16, 1855, in 7836 Fonds Enfantin; Lesseps to Madame Delamalle, Jan. 22, 1855, in Lesseps, *Lettres 1854–1856,* pp. 108–11.

11. Enfantin to Lesseps, Feb. 10, 1855, in 7835 Fonds Enfantin.

12. Enfantin to Negrelli, April 16, 1855, in 7836 Fonds Enfantin; Enfantin to M. Perron, May 4, 1855, in 7837 Fonds Enfantin.

13. Enfantin to M. Garbeiron, June 1, 1855, in 7837 Fonds Enfantin; Lesseps to Arlès-Dufour, June 18, 1855, in 7837 Fonds Enfantin.

CHAPTER EIGHT: THE SULTAN'S SHADOW AND THE ENGLISH LION

1. Russell quotation in Herbert C. F. Bell, *Lord Palmerston* (London: Longmans, Green, and Co., 1936), vol. 2, p. 424. For more on Palmerston, see Jaspar Ridley, *Lord Palmerston;* David Thomson, *England in the Nineteenth Century* (New York: Penguin, 1978); Charles Webster, *The Foreign Policy of Lord Palmerston* (London: Bell & Sons, 1951); Roy Jenkins, *Gladstone* (New York: Random House, 1997); Evelyn Ashley, *The Life and Correspondences of Viscount Palmerston;* E. D. Steele, *Palmerston and Liberalism, 1855–1865* (Cambridge, Eng.: Cambridge University Press, 1991); Andrew Porter, editor, *The Oxford History of the British Empire: The Nineteenth Century* (New York: Oxford University Press, 1998). Much of the more recent scholarship on Palmerston has emended the picture of Palmerston as a gruff jingoist. Revisionism is part and parcel of the writing of history, but while there was far more to England than the attitudes of Palmerston, the earlier picture of him as a staunch imperialist seems more credible to me than recent suggestions that he was a subtle leader enmeshed in a complicated society. Yes, mid-nineteenth-century Britain was a multifaceted society, and recent academic work has demonstrated the complexities. Palmerston remains, in my opinion, a rather straightforward figure.

2. D. K. Fieldhouse, *Economics and Empire, 1830–1914* (London: Macmillan, 1984); A. P. Thornton, *The Imperial Idea and Its Enemies* (London: Macmillan, 1985); Ronald Robinson and John Gallagher, *Africa and the Victorians* (London: Macmillion, 1961); Bernard Porter, *The Lion's Share: A Short History of British Imperialism 1850–1983* (London: Longman, 1984); David Cannadine, *Ornamentalism: How the British Saw Their Empire* (New York: Oxford University Press, 2001); Lawrence James, *The Rise and Fall of the British Empire* (New York: St. Martin's Press, 1994).

3. Queen Victoria quoted in *The Times,* May 2, 1851; Matthew Arnold, *Culture and Anarchy* (London, 1869).

4. Bruce to Lord Clarendon (Foreign Secretary), Dec. 3, 1854, and Lord Stratford, dispatch to the Foreign Secretary, Jan. 11, 1855, both in 78/1156 PRO.

5. Lesseps, "Note pour le vice-roi . . . ," Feb. 15, 1855, and letters to Lord Stratford, Feb. 26 and 28, 1855, in Ferdinand de Lesseps, *Lettres 1854–1856,* pp. 117–21, 127–28, 134–38; Stratford to Foreign Office, Feb. 22 and 26, 1855, in FO 78/1156 PRO.

6. Lesseps to M. Hippolyte Lafosse, March 22, 1855, in Lesseps, *Lettres 1854–1856,* pp. 155–61.

7. Said to Kiamil Pasha, March 31, 1855, in FO 78/1156 PRO.

8. Lesseps to Madame Delamalle, April 21, 1855, in Ferdinand de Lesseps, *Recollections of Forty Years,* p. 260.

9. Paulin Talabot, "Le Canal de deux mers," and J. J. Baude, "De l'isthme de Suez and du Canal Maritime à ouvrir de la Méditerranée et la Mer Rouge," *Revue des deux mondes,* March 15 and April 30, 1855.

10. Lesseps, note to Said, April 18, 1855; note to Count Walewski, June 7, 1855; and note to the emperor, June 9, 1855, all in Lesseps, *Lettres 1854–1856*, pp. 184–99.

### CHAPTER NINE: HITHER AND YON

1. Ferdinand de Lesseps to Count de Lesseps, June 25, 1855, in Ferdinand de Lesseps, *Recollections of Forty Years*, p. 269, and in Ferdinand de Lesseps, *Lettres 1854–1856*, pp. 221–27.

2. Lord Cowley, memorandum, July 2, 1855, in FO 78/1156 PRO.

3. *Athenaeum*, Aug. 25, 1855.

4. Lesseps to Emperor Napoleon, July 4, 1855, in Lesseps, *Lettres 1854–1856*, pp. 235–38.

5. Lesseps, statement on behalf of the viceroy to the members of the international commission, Dec. 16, 1855, in Lesseps, *Lettres 1854–1856*, pp. 319–20; see also Consul Green to Foreign Secretary, Jan. 6, 1856, in FO 78/1340 PRO. Dispatches of the commission are in 06892 CAMT.

6. Second Act of Concession, Jan. 5, 1856; Said, Decree as to the Native Workmen, July 20, 1856; and Statutes of the Company, all in Percy Hetherington Fitzgerald, *The Great Canal at Suez*, pp. 297–323.

7. Enfantin to Gabeiron, Aug. 4, 1855; Enfantin to the emperor, Sept. 10 and Oct. 24, 1855; official reply to Enfantin, Dec. 15, 1855, all in 7836 Fonds Enfantin. On Stephenson, see Dan Bradshaw, "Stephenson, de Lesseps, and the Suez Canal," *Journal of Transport History*, Sept. 1978.

8. Lesseps to Barthélemy Saint-Hillaire, April 7, 1856, in Lesseps, *Lettres 1854–1856*, pp. 377–79. Report of Newcastle meeting, May 28, 1857, in FO 78/1340 PRO. Reports of other meetings in Ferdinand de Lesseps, *Lettres, 1857–1858*, pp. 72–87.

9. The exchanges between Lesseps and Palmerston and Lesseps and Stephenson can be found in Lesseps, *Lettres 1857–1858*, pp. 87–113; some of those letters are also translated in Lesseps, *Recollections*, pp. 55–69. Speeches of Palmerston and Stephenson in the House of Commons, July 7 and July 17, 1857, in *Hansard's Parliamentary Debates*.

10. Lord Cowley, dispatch, April 3, 1856, in FO 78/1340 PRO.

11. John Freely, *Istanbul: The Imperial City* (New York: Penguin, 1996), pp. 221–80; Lord Kinross, *The Ottoman Centuries: The Rise and Fall of the Turkish Empire* (New York: William Morrow, 1977), pp. 417–507; Bernard Lewis, *The Emergence of Modern Turkey* (New York: Oxford University Press, 1961); Sanford Shaw and Ezel Kural Shaw, *History of the Ottoman Empire and Modern Turkey*, vol. 2 (Cambridge, Eng.: Cambridge University Press, 1977).

### CHAPTER TEN: THE EMPEROR AND THE ENTREPRENEUR

1. The quotation about Napoleon as a man of one idea comes from Philip Guedella, *The Second Empire* (New York: G. Putnam, 1922), p. 243. The details about

Napoleon come from multiple sources: John Bierman, *Napoleon III and His Carnival Empire* (New York: St. Martin's, 1988); Fenton Bressler, *Napoleon III: A Life* (New York: Carroll & Graf, 1999); David Duff, *Eugenie and Napoleon III: A Dual Biography* (New York: William Morrow, 1978); Louis Girard, *Napoleon III* (Paris: Libraire Arthème Fayard, 1986); Pierre Miquel, *Le Second Empire;* Alain Plessis, *De la fête impériale au mur des fédérés 1852–1871* (Paris: Éditions du Seuil, 1979); Robert Tombs, *France: 1814–1914* (London: Longman, 1996); Roger Williams, *The World of Napoleon III, 1851–1870* (New York: Free Press, 1957); Theodore Zeldin, *France 1848–1945: Politics and Anger;* Roger Price, *The French Second Empire: An Anatomy of Power* (Cambridge, Eng.: Cambridge University Press, 2001); Philip Mansel, *Paris Between Empires, 1814–1852* (London: John Murray, 2001).

2. The meeting between Ali and the emperor was discussed by Lesseps in a letter to M. Thouvenal, the French ambassador in Constantinople, April 22, 1856, in Ferdinand de Lesseps, *Recollections of Forty Years,* pp. 294–95. The benediction is from Lesseps, letter of March 21, 1856, *Lettres 1854–1856,* pp. 351–52, quoted in Lord Kinross, *Between Two Seas,* p. 95.

3. Cowley to Clarendon, Dec. 26, 1856, in FO 78/1340 PRO.

4. Quoted in multiple sources, including Ghislain de Diesbach, *Ferdinand de Lesseps,* pp. 161–62, and Charles Beatty, *De Lesseps of Suez,* p. 175.

5. Bierman, *Napoleon III,* pp. 175–90; Duff, *Eugénie and Napoleon III,* pp. 126–38; Herbert C. F. Bell, *Lord Palmerston,* pp. 180–83.

6. Lesseps to Le Comte de Lesseps, and Lesseps to Negrelli, both April 17, 1858, in Ferdinand de Lesseps, *Lettres 1857–1858,* pp. 195–203.

7. Report, May 10, 1858, in 1522 CAMT.

8. Debate in the House of Commons can be found both in *Hansard's Parliamentary Debates* and in *The Times,* June 2, 1858. For an example of the many British defenders of the canal, see "Suez Ship Canal," *Dublin University Magazine,* May 1858, as well as editorials in *Daily News,* June 3, 1858. See also "Egypt and the Suez Canal," *Fraser's Magazine,* Feb. 1860. For the continued opposition of Stephenson, see a long letter he wrote to *The Times,* Aug. 3, 1858. Lesseps quoted in George Edgar-Bonnet, *Ferdinand de Lesseps,* p. 301.

CHAPTER ELEVEN: A UNIVERSAL COMPANY

FOR A MARITIME CANAL

1. Lesseps circulars, in Ferdinand de Lesseps *Lettres 1857–1858,* pp. 352–58. The circular stated that 150 francs would be due once the allocation had been settled, but it appears that this was subsequently amended to fifty francs.

2. Account of Odessa banquet taken from *Journal d'Odessa,* Aug. 9, 1858, in Lesseps, *Lettres 1857–1858,* p. 313. Account of Barcelona meeting taken from *Diario,* Oct. 1858, in Lesseps, *Lettres 1857–1858,* pp. 360–65; account of Turin meeting from Lesseps, *Lettres 1857–1858,* pp. 372–73.

3. Preliminary division of shares, Sept. 1858, in 1164 CAMT. List of shareholders in Lesseps, *Lettres 1857–1858,* pp. 380–88; list of Egyptian shareholders, in 1164

CAMT. See also Werner Baer, "The Promoting and Financing of the Suez Canal," *Business History Review,* Fall 1956, pp. 361–68.

4. Lesseps to Baron Bruck, Dec. 18, 1858; Lesseps to M. Revoltella, April 21, 1859; Revoltella to Lesseps, all in 1164 CAMT. List of shareholders, according to the company's accounts, included fifty thousand shareholders from Austria, twenty-four thousand from Russia, and five thousand each from the United States and Great Britain. There is some debate about whether a handful of these shares were actually bought in those countries. For the list, see *L'Isthme de Suez,* Jan. 19, 1859, in 1521 CAMT. For the debate, see George Edgar-Bonnet, *Ferdinand de Lesseps,* p. 327.

5. Said quoted in De Regny to Lesseps, Jan. 2, 1859, in Ferdinand de Lesseps, *Lettres 1859–1860,* pp. 16–17. For other accounts of the interaction between Said and Lesseps, see David Landes, *Bankers and Pashas,* pp. 174ff; Robert Solé, *L'Égypte, passion française,* pp. 178–79; George Edgar-Bonnet, *Ferdinand de Lesseps,* pp. 315–30; Nubar Pacha, *Mémoires de Nubar Pacha,* pp. 140–90.

6. Notes from the first meeting of the Conseil d'Administration, Dec. 20, 1858, in 1164 CAMT.

7. *The Times,* Nov. 27, 1858; other newspaper quote in Edgar-Bonnet, *Lesseps,* p. 327; Palmerston quoted in Lord Kinross, *Between Two Seas,* p. 115.

8. Once again, the issue of how the West depicted the East has been a source of intense controversy in academic and intellectual circles in Europe and the United States in the past three decades, especially since the publication of Edward Said's *Orientalism.*

9. For a discussion of art, see *Picturing the Middle East: A Hundred Years of European Orientalism—A Symposium* (New York: Dahesh Museum, 1996); Rana Kabbani, *Europe's Myths of Orient* (London: Pandora Press, 1986); Christine Peltre, *Orientalism in Art* (New York: Abbeville Press, 1998); Arthur Danto, "The Late Works of Delacroix," *Nation,* Nov. 9, 1998; Anita Brookner, *Romanticism and Its Discontents,* pp. 80–119.

10. On Flaubert, see Jean-Marie Carré, *Voyageurs et écrivains français en Égypte,* pp. 83–130; Gustave Flaubert, *Flaubert in Egypt,* ed. and trans. Francis Steegmuller (New York: Penguin, 1972); Mary Orr, "Flaubert's Egypt," in Paul Starkey and Janet Starkey, eds., *Travellers in Egypt,* pp. 189–200; Geoffrey Wall, *Flaubert* (New York: Farrar, Straus & Giroux, 2002).

11. This point is very well made in Ruth Bernard Yeazell, *Harems of the Mind: Passages of Western Art and Literature* (New Haven: Yale University Press, 2000).

12. Giles Lambert, *Auguste Mariette: L'Égypte ancienne sauvée des sables* (Paris: J. C. Lattes, 1997).

### CHAPTER TWELVE: THE WORK AHEAD

1. Press clipping in *L'Isthme de Suez,* Nov. 25, 1858, in 1521 CAMT. Lesseps to Said, Dec. 31, 1858, in Ferdinand de Lesseps, *Lettres 1857–1858,* pp. 408–12; similar but slightly different version, dated Jan. 5, 1859, in 0002 CAMT. Various correspondence between Lesseps and shareholders, in 1164 CAMT.

2. Minutes of the Conseil Supérieur des Travaux, Nov. 1858, in 0692 CAMT.

3. Agreement between the company and Hardon, Feb. 12, 1859, in Ferdinand de Lesseps, *Lettres 1859–1860*, pp. 20–27. Nathalie Montel, *Le Chantier du Canal de Suez*, pp. 30–35. Minutes of the General Assembly of Shareholders, with documents, May 15, 1860, in 0002 CAMT.

4. Various correspondence between Consul Green and foreign secretary, Feb. 3 and March 7, 1859, and between Acting Consul Walne and foreign secretary, April 14, May 6, May 25, and June 3, 1859, all in FO 78/1489 PRO. Lesseps to Said, April 9 and June 8, 1859, in Lesseps, *Lettres 1859–1860*, pp. 67–70, 131–33.

5. Grand Vizier Ali Pasha to Said, c. June 20, 1859, in Lesseps, *Lettres 1859–1860*, pp. 158–59; Acting Consul Walne to Bulwer, July 6, 1859, in FO 78/1489 PRO.

6. Lesseps to Ruyssenaers, Oct. 24, 1859, in Lesseps, *Lettres 1859–1860*, pp. 235–39.

7. Report of Consul Colquhoun, April 26, 1860, in FO 78/1556 PRO. Engineer quoted in Lord Kinross, *Between Two Seas*, p. 146. *The Times*, May 23 and 24, 1860. For accounts of the engineering challenges, see Jean-Paul Calon, "The Suez Canal Revisted: Ferdinand de Lesseps, the Genesis and Nurturing of Macroengineering Projects for the Next Century," *Interdisciplinary Science Reviews*, vol. 19, no. 3 (1994); D. F. Bradshaw, "A Decade of British Opposition to the Suez Canal Project, 1854–1864," *Transport History*, Spring 1978; J. Clerk, "Suez Canal," *Fortnightly Review*, Jan. 1869.

8. Account taken from *L'Isthme de Suez*, June 15, 1860, in 1521 CAMT; also from George Edgar-Bonnet, *Ferdinand de Lesseps*, pp. 459–62; Ghislain de Diesbach, *Ferdinand de Lesseps*, pp. 183–86.

9. Palmerston, address to the House of Commons, Aug. 23, 1860, in *Hansard's*, pp. 1723–24.

10. Montel, *Chantier*, p. 41; Pudney, *Suez*, p. 95.

11. Voisin left an extensive record of his years with the company. See Voisin Bey, *Le Canal de Suez*, 7 vols. (Paris: Charles Dunod, 1902–6). He also left a voluminous collection of papers, which are housed at CAMT. See also Jean-Édouard Goby, "Un Grand Ingénieur français: Voisin Bey," unpublished paper, 1958, in 1180 CAMT.

12. Montel, *Chantier*, p. 41; Edgar-Bonnet, *Lesseps*, pp. 452–53; John Marlowe, *World Ditch*, pp. 134–35.

## CHAPTER THIRTEEN: THE CORVÉE

1. On the corvée and the army under Muhammad Ali, see Afaf Lufti al-Sayyid Marsot, *Egypt in the Reign of Muhammad Ali*, pp. 109, 121, 150–51; Khaled Fahmy, *All the Pasha's Men*; Fahmy, "The Era of Muhammad Ali Pasha," *The Cambridge History of Egypt*. See also Lawrence Jennings, *French Anti-Slavery: The Movement for the Abolition of Slavery in France, 1802–1848* (New York: Cambridge University Press, 2000).

2. Lucy Duff-Gordon quoted in Katherine Frank, *A Passage to Egypt: The Life of Lucie Duff Gordon* (Boston: Houghton Mifflin, 1994), p. 249. On Sufism, see

Annemarie Schimmel, *Mystical Dimensions of Islam* (Chapel Hill: University of North Carolina Press, 1975); Ira Lapidus, *A History of Islamic Societies* (Cambridge, Eng.: Cambridge University Press, 1988), pp. 260–65, 359–65, 615–21; J. S. Trimingham, *The Sufi Orders in Islam* (Oxford: Clarendon Press, 1971). On social history, see Gabriel Baer, "Continuity and Change in Egyptian Rural Society, 1805–1882," in Groupe de Recherches et d'Études sur le Proche-Orient, *L'Égypte au XIX siècle*, pp. 231–37; Ehud Toledano, "Social and Economic Change in the Long Nineteenth Century," in *The Cambridge History of Egypt*, vol. 2, pp. 252–84; Timothy Mitchell, *Colonising Egypt* (New York: Cambridge University Press, 1988).

3. Lesseps quotation in Lord Kinross, *Between Two Seas*, p. 148. For public-health records, see "Report of the Health Service," *L'Isthme de Suez*, May 1, 1861, in 1521 CAMT. On treatment of workers, see report of Henry Bulwer, Jan. 1, 1863, in FO 78/1795 PRO; J. Clerk, "Suez Canal," pp. 206–7.

4. Statements of Darby Griffith in the House of Commons, June 21, 1861, and May 16, 1862, in *Hansard's*.

5. Ghislain de Diesbach, *Ferdinand de Lesseps*, pp. 200–201; J. E. Nourse, *The Maritime Canal of Suez*, pp. 24–25; statement of the mufti of Cairo, in 1172 CAMT.

### CHAPTER FOURTEEN: THE NEW VICEROY AND HIS MINISTER

1. "Ismail: Pacha of Egypt," *Harper's Magazine*, no. 39, 1869, pp. 739–41; Pierre Crabites, *Ismail*, pp. 31–32; "Mort de S. A. Muhammad-Said Pacha," *L'Isthme de Suez*, Feb. 1, 1863, in 1521 CAMT.

2. Bulwer, report, Jan. 4, 1863, in FO 78/1795 PRO. Roger Owen, *The Middle East in the World Economy, 1800–1914*, pp. 122–29; David Landes, *Bankers and Pashas*, pp. 128ff; Crabites, *Ismail*, p. 144; F. Robert Hunter, "Egypt Under the Successors of Muhammad Ali," in *The Cambridge History of Egypt*, vol. 2, pp. 180–190; Emine Foat Tugay, *Three Centuries*, pp. 125–37; on Ismail's attitude towards Europe, see P. J. Vatikiotis, *The History of Egypt*, pp. 70–89. Abdin Palace still sits in Cairo and is still used as the center of government. Various knickknacks and artifacts from Ismail's palace and from the army of the time can be seen in the Military Museum at the Citadel in Cairo, and at the National Museum at Port Said.

3. Ismail to assembled consuls quoted in G. Douin, *Histoire du règne du Khédive Ismail*, vol. 1, *Les Premières Années du règne* (Rome: Istituto Poligrafico dello Stato, 1933), pp. 2–3. Press reports found in J. E. Nourse, *The Maritime Canal of Suez*, p. 29. The statement about Ismail's being more *canaliste* than Lesseps has been widely quoted, but there are slight variations in the phrasing. This one is taken from Crabites, *Ismail*, p. 45. See also John Marlowe, *World Ditch*, p. 155. Ismail's later statements on the corvée found in Colquhoun to Foreign Office, Jan. 23 and 24, 1863, in FO 78/1795 PRO.

4. Quotation about Napoleon from George Edgar-Bonnet, *Ferdinand de Lesseps*, p. 371. On the British stance, see K. Bell, "British Policy Towards the Construction of the Suez Canal, 1859–1865," *Transactions of the Royal Historical Society*, pp. 121–43.

On Abdul Aziz's visit, see Vatikiotis, *History of Egypt,* pp. 74–75; Douin, *Histoire,* vol. 1, pp. 9–17.

5. Nubar Pacha, *Mémoires de Nubar Pacha;* Mirit Boutros-Ghali, "Les Mémoires de Nubar Pacha," in Groupe de Recherches et d'Études sur le Proche-Orient, *L'Égypte au XIX siècle,* pp. 35–47; F. Robert Hunter, *Egypt Under the Khedives, 1805–1879,* pp. 165–74; Edgar-Bonnet, *Lesseps,* pp. 371–77.

CHAPTER FIFTEEN: FERDINAND FIGHTS BACK

1. Lesseps to General Assembly, July 15, 1863, quoted in George Edgar-Bonnet, *Ferdinand de Lesseps,* pp. 381–82. Lesseps to Théodore de Lesseps, Aug. 28, 1863, in Ferdinand de Lesseps, *Lettres 1863–1864,* pp. 326–27.

2. Court papers against Nubar, in 1165 CAMT. Lesseps to the Council, Oct. 13, 1863, in Edgar-Bonnet, *Lesseps,* p. 388.

3. Nubar on Lesseps and story of the shareholder approaching Nubar in Nubar Pacha, *Mémoires de Nubar Pacha,* pp. 223–28. For the exchange between Nubar and Lesseps, see also Nubar, letter, Oct. 14, 1863, and Lesseps's reply, in Lesseps, *Lettres 1863–1864,* pp. 388–92. This scene is nicely paraphrased in Lord Kinross, *Between Two Seas* p. 185.

4. Émile Ollivier to Duc de Morny, Jan. 16, 1864, in 1165 CAMT.

5. Lesseps to emperor, Jan. 6, 1864, quoted in Charles Beatty, *De Lesseps of Suez,* p. 232. Though Lesseps refers to 1860, the emperor's crucial support had actually come at the end of 1859; see page 160.

6. Quoted in Edgar-Bonnet, *Lesseps,* p. 399.

7. Report, Nubar to Morny, "Question des corvées," March 1864, in 1165 CAMT. Nubar, letters to Madama Nubar, in Nubar, *Mémoires,* pp. 233–34. Palmerston to House of Commons, in *The Times,* April 12, 1864.

8. Used as a piece of propaganda for the company, the meeting was summarized and the speeches were reprinted in *L'Isthme de Suez,* Feb. 15, 1864; inauguration of Sweet Water Canal reported in ibid., Jan. 15, 1864, both in 1521 CAMT.

9. The voluminous records of the arbitration commission are in 1165 CAMT.

10. Quote from Nubar, *Mémoires,* p. 239. Text of the arbitration (Sentence Arbitrale), July 6, 1854, in 0725 CAMT.

11. David Landes, *Bankers and Pashas,* pp. 189ff; G. Douin, *Histoire du règne du Khédive Ismail,* vol. 1, p. 141; Crabites, *Ismail,* pp. 30–40.

12. Maxime Du Camp, *Souvenirs littéraires,* p. 103, quoted in Ghislain de Diesbach, *Ferdinand de Lesseps,* p. 148.

CHAPTER SIXTEEN: MEN AND MACHINES

1. Digest of negotiations prepared by the company's Conseil d'Administration, in 0633 CAMT.

2. Nathalie Montel, *Le Chantier du Canal de Suez,* pp. 72–73, 265–70.

3. Ibid., pp. 231ff; Tom Peters, *Building the Nineteenth Century* (Cambridge, Mass.: MIT Press, 1996), pp. 178–201; J. E. Nourse, *The Maritime Canal of Suez*, pp. 31–38; J. Clerk, "Suez Canal," pp. 88–95; "Chronique de l'isthme," *L'Isthme de Suez*, Sept. 15, 1864, in 1521 CAMT; minutes from the Annual General Assembly of Shareholders, Oct. 5, 1866, *L'Isthme de Suez*, Oct. 12–15, 1866, in 1520 CAMT. Cost per cubic meter found in various work contracts in the Voisin papers, in 1182 CAMT. Each annual shareholders' meeting included an extensive report on the status of the work. See also Lavalley, report to Civil Society of Engineers, Sept. 21, 1866, *L'Isthme de Suez*, Oct. 15, 1866, in 1523 CAMT.

4. Lesseps, report to shareholders, Aug. 1, 1866, in 0022 CAMT.

5. Comments about the fishermen in Lesseps, speech in Lyon, Feb. 12, 1865, *L'Isthme de Suez*, March 1, 1865, in 1523 CAMT. See also Laroche to Voisin, April 11, 1864; Voisin to Gerardin, April 19, 1864; Lesseps to Girardin, May 1, 1864, all in 3297 CAMT.

6. "Les Délégations commerciales dans L'Isthme de Suez," *L'Isthme de Suez*, May 1, 1865, in 1523 CAMT.

7. "Le Bilan du cholera," *L'Isthme de Suez*, Aug. 15, 1865, in 1523 CAMT. Press clippings from *Journal de travaux publics* and subsequent court papers, also in 1523 CAMT. For the history of cholera, see Charles Rosenberg, *The Cholera Years* (Chicago: University of Chicago Press, 1962); Norman Longmate, *King Cholera* (London: Hamish Hamilton, 1966); Jacques Barzun, *From Dawn to Decadence: 1500 to the Present* (New York: HarperCollins, 2000), p. 497.

8. "Lord Palmerston," *L'Isthme de Suez*, Nov. 1, 1865.

9. Quoted in F. Robert Hunter, *Egypt Under the Khedives, 1805–1879*, p. 35.

10. Text of the agreement between the company and the Egyptian government, Feb. 22, 1866, and text of the sultan's firman, March 19, 1866, in Percy Hetherington Fitzgerald, *The Great Canal at Suez*, pp. 329–33.

### CHAPTER SEVENTEEN: THE CANAL GOES TO PARIS

1. "Rapport de M. Ferdinand de Lesseps," minutes, General Assembly of Shareholders, Aug. 1, 1867, in 0022 CAMT; "Assemblée des actionnaires," *La Presse*, Aug. 5, 1867, in 1018 CAMT.

2. Blunt quoted in John Bierman, *Napoleon III and His Carnival Empire*, p. 259. Other sources for the exposition include, Pierre Miquel, *Le Second Empire*, pp. 376–404; Annie Cohen-Solal, *Painting American: The Rise of American Artists, Paris 1867–New York 1948* (New York: Alfred A. Knopf, 2001), pp. 3–63; Patricia Mainardi, *Arts and Politics of the Second Empire: The Universal Expositions of 1855 and 1867* (New Haven: Yale University Press, 1987); Victor Hugo quoted in Arthur Chandler, "The Paris *Exposition Universelle* of 1867," *World's Fair Magazine*, no. 3, 1986; Alistair Horne, *Seven Ages of Paris* (New York: Knopf, 2002), pp. 245–47. On the culture of industrialization in France, see Jean-Pierre Daviet, *La Société industrielle en France* (Paris: Éditions du Seuil, 1997).

3. "L'Égypte à l'Exposition Universelle de 1867," *L'Isthme de Suez*, April 1, 1866,

and "Autour de l'Exposition Universelle," *L'Isthme de Suez*, May 1, 1867, both in 1523 CAMT; "Exposition Universelle: L'Isthme de Suez," *Journal des débats*, July 22, 1867, in 1018 CAMT; "L'Égypte," *Le Monde*, Nov. 13, 1867, in 1019 CAMT; Robert Solé, *L'Égypte, passion française*, pp. 208–15.

4. Description of panorama in *Le Figaro*, June 8, 1867. Théophile Gautier, in *Le Globe artiste et industriel*, Oct. 8, 1867, in 1019 CAMT.

5. H. Vrignault, "Cent soixante-dix kilomètres," *La Presse*, June 7, 1867; "Inauguration de l'exposition," *Le Soleil*, June 6, 1867; "L'Isthme de Suez à l'exposition," *Le Siècle*, July 29, 1867, all in 1018 CAMT.

6. Giles Lambert, *Auguste Mariette*, p. 219.

7. *Standard*, Sept. 28, 1867; *Money Market Review*, Sept. 28, 1867, in 1019 CAMT. Critical reactions in the French press found in a variety of clippings dating from late Sept. and early Oct., in 1018 CAMT.

8. The transcripts of the Legislative Corps debates are in 1166 CAMT. Quotation from *Journal de Paris*, Sept. 25, 1867, in 1018 CAMT.

9. "Les Autres Lesseps," *Paris Magazine*, Jan. 13, 1867, in 1017 CAMT; "Ferdinand de Lesseps," *Journal de Nice*, Dec. 1, 1867, in 1019 CAMT.

10. Roger Owen, *The Middle East in the World Economy, 1800–1914*, pp. 150ff; David Landes, *Bankers and Pashas*, pp. 173ff. On Ismail's reception in France, see Solé, *L'Égypte*, pp. 216–28; "Ismail en France," *L'Isthme de Suez*, June 15, 1867, in 1523 CAMT; Nubar quotation in Nubar Pacha, *Mémoires de Nubar Pacha*, pp. 311–12.

## CHAPTER EIGHTEEN: THE FINAL STAGES

1. Each month, a detailed status report was published by the company and then printed in the company's journal and in many of the leading French papers. Though the company certainly had an interest in presenting its progress in the best light, it could not publish blatantly false data. Too many experts were visiting the canal zone, and they would have been able to tell if there was a significant discrepancy between what the company said and what was actually happening on the ground. See, for instance, "Situation générale des travaux à la fin du mois de décembre 1867," *L'Isthme de Suez*, Jan. 15–18, 1868, in 1523 CAMT.

2. Quoted in Ghislain de Diesbach, *Ferdinand de Lesseps*, p. 238. Population figures from *L'Isthme de Suez*, May 15, 1868, in 1523 CAMT.

3. For a sample of European press coverage on the judicial disputes, see multiple clippings in 1020 CAMT, such as "La Réforme judiciare en Égypte," *La Presse*, Sept. 13, 1868; "La Jurisdiction consulaire en Égypte," *L'Opinion nationale*, Oct. 1, 1868; "Des Tribunaux mixtes en Orient," *La Patrie*, Oct. 2, 1868.

4. Editorial in *La Terre sainte*, Jan. 23, 1868, in 1019 CAMT. Article on the courts as a "school" for Egypt, in *L'Opinion nationale*, Oct. 12, 1868, in 1020 CAMT.

5. The customs-duty dispute is extensively documented in 3599 CAMT.

6. *L'Isthme de Suez*, May 15, 1867, in 1523 CAMT; estimates of tonnage in Lesseps, report to shareholders, Aug. 1868, in 0023 CAMT.

7. L. Alloury, "Panorama," *Le Courrier français*, June 12, 1868, in 1021 CAMT.

8. *The Times,* Dec. 29, 1868, and Feb. 18, 1869. Riou quoted in Pudney, *Suez,* p. 139.

9. Quote about Moses in Diesbach, *Lesseps,* p. 251. The letter from Ismail to Lesseps, April 23, 1869, is widely quoted in a number of books; the version quoted is my translation from a copy of the original, marked *"confidentielle,"* found in 1172 CAMT.

10. *The Times,* April 12, 1869. Quotation about Palmerston in Charles Beatty, *De Lesseps of Suez,* p. 249.

11. June Hargrove, "Liberty: Bartholdi's Quest for a Visual Metaphor," *The World & I,* July 1986; Marvin Trachtenberg, *The Statue of Liberty* (New York: Penguin, 1986); Ahmed Yousseff, *La Fascination de l'Égypte,* pp. 411–12.

12. Various issues of *L'Isthme de Suez* in 1523 CAMT.

### CHAPTER NINETEEN: THE DESERT IS PARTED

1. Breakdown of the expenses for the festival in 1025 CAMT.

2. Quoted in Christophe Pincemaille, *L'Impératrice Eugénie: De Suez à Sedan* (Paris: Payot, 2000), p. 107.

3. Lists of invitations in 1734 CAMT. "Rapport: Présénté à l'Assemblée générale du 2 août 1869," in 1023 CAMT. Assorted French press clippings and publicity about Lesseps and the opening, July and Aug. 1869, also in 1023 CAMT. Advertisements for excursions, in 1024 CAMT. *Daily Telegraph,* Aug. 26, 1869.

4. *L'Isthme de Suez,* Aug. 18, 1869, in 1523 CAMT.

5. Pincemaille, *L'Impératrice Eugénie,* pp. 163ff; Robert Solé, *L'Égypte, passion française,* pp. 229–36.

6. Jean-Marie Carré, *Voyageurs et écrivains français en Égypte,* pp. 200–209, 305–8.

7. Quoted in Daniel Boorstin, *The Creators: A History of Heroes of the Imagination* (New York: Random House, 1992), p. 474.

8. Much of this chapter relies on more than a thousand articles and clippings in 1024 CAMT from Oct. and Nov. 1869. Reports about the preparations can be found in leading French papers such as *Le Temps, Le Réveil, Le Figaro, Le Moniteur universel,* and *Le Siècle,* and there is a clippings file dedicated to the opening ceremonies, in 0723 CAMT. Arrow quotation in Sir Frederick Arrow, *Fortnight in Egypt at the Opening of the Suez Canal* (London, 1869). See also "The Opening of the Suez Canal," *Blackwood's Magazine* (Jan.–Feb.–Mar. 1870); F. A. Eaton, "The Suez Canal," *Macmillan's Magazine,* Dec. 1869, pp. 82–95. Lengthy dispatches were published in *The Times,* Nov. 20 and Dec. 7, 1869. Nubar Pasha gave his account of the ceremonies, in Nubar Pacha, *Mémoires de Nubar Pacha,* pp. 359–65. There are also colorful descriptions in Lord Kinross, *Between Two Seas,* pp. 236ff; John Marlowe, *World Ditch,* pp. 223–34; and Ghislain de Diesbach, *Ferdinand de Lesseps,* pp. 262–71; these rely on a smaller set of newspapers or earlier secondary accounts, however.

9. Lesseps, *Lettres, 1864–1869. L'Isthme de Suez,* Dec. 15, 1869, in 1523 CAMT.

CHAPTER TWENTY: THE LEGACY

1. Roger Owen, *The Middle East in the World Economy, 1800–1914*, p. 127. Most of the information and quotations in this chapter can be found in a wide variety of secondary sources, all of which have been cited previously.

2. The single best account of the Panama Canal is David McCullough, *The Path Between the Seas* (New York: Simon & Schuster, 1977).

3. Quoted in Derek Hopwood, *Egypt: Politics and Society 1945–90* (London: Routledge, 1991), p. 48. See also the encyclopedic account of the Suez crisis in Keith Kyle, *Suez* (London: Weidenfeld & Nicolson, 1991); and, for facts about the modern canal, George Lenczowski, *The Middle East in World Affairs: Fourth Edition* (Ithaca, N.Y.: Cornell University Press, 1980).

# Select Bibliography

A *note on the sources*: Most of the narrative derives from primary sources. Ferdinand de Lesseps left several volumes of published letters, and major newspapers and journals in France, Great Britain, and throughout Europe reported on the canal, its politics, and its finances throughout this period. In addition, the Suez Canal Company archives are housed in Roubaix, France, at the Centre des Archives du Monde du Travail, and these were vital. Equally important were the records of the British Foreign Office for these years, which can be found at the Public Records Office in Kew, London. An underutilized source are the papers of Prosper Enfantin, which are located in the Arsenal Library in Paris and which cover not only the politics of the Saint-Simonians but the initial stages of the canal as well. There are very few Arabic sources, either in archives in Cairo or in published form, and with a notable exception or two, most of the secondary sources in Arabic rely on French and English accounts of the canal. The following is a list of the more important secondary sources in English and French that were consulted during the course of the research.

Adkins, Lesley and Roy. *The Keys of Egypt: The Race to Read the Hieroglyphs* (New York: HarperCollins, 2000).
Anderson, M. S. *The Eastern Question* (London: Macmillan, 1986).
Annesley, George. *The Rise of Modern Egypt: A Century and a Half of Egyptian History, 1798–1957* (Edinburgh: Pentland Press, 1994).
Ashley, Evelyn. *The Life and Correspondences of Viscount Palmerston* (London: Richard Bentley & Sons, 1879).
Asprey, Robert. *The Rise of Napoleon Bonaparte* (New York: Basic Books, 2000).
Barzun, Jacques. *From Dawn to Decadence: 1500 to the Present* (New York: HarperCollins, 2000).
Beatty, Charles. *De Lesseps of Suez: The Man and His Times* (New York: Harper and Brothers, 1956).
Bell, Herbert C. F. *Lord Palmerston* (London: Longmans, Green, and Co., 1936).
Bierman, John. *Napoleon III and His Carnival Empire* (New York: St. Martin's, 1988).
Bradford, Sarah. *Disraeli* (Briarcliff Manor: Stein & Day, 1982).

Brent, Peter. *Far Arabia: Explorers of the Myth* (London: Weidenfeld and Nicolson, 1977).

Bressler, Fenton. *Napoleon III: A Life* (New York: Carroll & Graf, 1999).

Brookner, Anita. *Romanticism and Its Discontents* (New York: Farrar, Straus, & Giroux, 2000).

Brown, Frederick. *Zola: A Life* (Baltimore: Johns Hopkins University Press, 1995).

Brown, L. Carl. *International Politics and the Middle East* (Princeton: Princeton University Press, 1984).

Cannadine, David. *Ornamentalism: How the British Saw Their Empire* (New York: Oxford University Press, 2001).

Cannon, Byron. *Politics of Law and the Courts in Nineteenth-Century Egypt* (Salt Lake City: University of Utah Press, 1988).

Carlisle, Robert. *The Proffered Crown: Saint-Simonianism and the Doctrine of Hope* (Baltimore: Johns Hopkins University Press, 1987).

Carré, Jean-Marie. *Voyageurs et Écrivains Français en Égypte* (Cairo: Institut Français d'Archéologie du Caire, 1956).

Castille, Hippolyte. *Le Père Enfantin* (Paris, 1859).

Charle, Christophe. *Histoire sociale de la France au XIX siècle* (Paris: Éditions du Seuil, 1991).

Charlton, D. G. *Secular Religions in France 1815–1870* (London: Oxford University Press, 1963).

Cohen-Solal, Annie. *Painting American: The Rise of American Artists, Paris 1867–New York 1948* (New York: Knopf, 2001).

Cooper, Duff. *Talleyrand* (New York: Grove Atlantic, 2001).

Crabbs, Jack. *The Writing of History in Nineteenth-Century Egypt: A Study in National Transformation* (The American University in Cairo Press, 1984).

Crabites, Pierre. *Ismail: The Maligned Khedive* (London: George Routledge & Sons, 1933).

Daly, M. W., ed. *The Cambridge History of Egypt: Volume Two* (New York: Cambridge University Press, 1998).

Daviet, Jean-Pierre. *La Société industrielle en France* (Paris: Éditions de Seuil, 1997).

de Diesbach, Ghislain. *Ferdinand de Lesseps* (Paris: Perrin, 1998).

Démier, Francis. *La France du XIX siècle* (Paris: Éditions de Seuil, 2000).

Dewachter, Michel, and Alain Fouchard, eds. *L'Égyptologie et les Champollion* (Grenoble: Presses Universitaires de Grenoble, 1994).

Dicey, Edward. *The Story of the Khedivate* (London: Rivingtons, 1902).

Dodwell, Henry. *The Founder of Modern Egypt* (Cambridge, Eng.: Cambridge University Press, 1931).

Douin, G. *Histoire du règne du Khédive Ismail, Vol. I: Les premières années du règne* (Rome: Instituto Poligrafico dello stato, 1933):

Duff, David. *Eugénie and Napoleon III: A Dual Biography* (New York: Morrow, 1978).

Edgar-Bonnet, George. *Ferdinand de Lesseps: Le Diplomate, Le Créator de Suez* (Paris: Librarie Plon, 1951).

Enfantin, Barthélemy-Prosper. *Life Eternal: Past-Present-Future,* translated by Fred Rothwell (Chicago: The Open Court Publishing Company, 1920).

Fahmy, Khaled. *All the Pasha's Men: Mehmed Ali, His Army, and the Making of Modern Egypt* (Cambridge, Eng.: Cambridge University Press, 1997).

Feiler, Bruce. *Walking the Bible* (New York: Morrow, 2001).

Ferguson, Niall. *The House of Rothschild: Money's Prophets 1798–1848* (New York: Viking, 1998).

Fieldhouse, D. K. *Economics and Empire, 1830–1914* (London: Macmillan, 1984).

Forester, E. M. *Alexandria* (New York: Oxford University Press, 1986).

Frank, Katherine. *A Passage to Egypt: The Life of Lucie Duff Gordon* (Boston: Houghton Mifflin, 1994).

Freely, John. *Istanbul: The Imperial City* (New York: Penguin, 1996).

Girard, Louis. *Napoleon III* (Paris: Librairie Arthème Fayard, 1986).

Goubert, Pierre. *Louis XIV and Twenty-Million Frenchmen,* translated by Ann Carter (New York: Vintage, 1966).

Gran, Peter. *Islamic Roots of Capitalism: Egypt 1760–1840* (Austin: University of Texas Press, 1979).

Groupe de Recherches et d'Études sur le Proche-Orient, *L'Égypte au XIX Siècle* (Paris: Éditions du Centre national de la recherche scientifique, 1982).

Guedella, Philip. *The Second Empire* (New York: G. Putnam, 1922).

Harrison, Robert. *Gladstone's Imperialism in Egypt: Techniques of Domination* (Westport, Conn.: Greenwood Press, 1995).

Hobsbawm, E. J. *The Age of Revolution 1789–1848* (New York: New American Library, 1962).

Hopwood, Derek. *Egypt: Politics and Society 1945–90* (London: Routledge, 1991).

Hourani, Albert. *Arabic Thought in the Liberal Age, 1798–1939* (Cambridge: Cambridge University Press, 1983).

———. *A History of the Arab Peoples* (Cambridge, Mass.: Harvard University Press, 1991).

———. *A Vision of History* (Beirut, 1961).

Hunter, F. Robert. *Egypt Under the Khedives, 1805–1879* (University of Pittsburgh Press, 1984).

Iggers, Georg, trans. *The Saint-Simon: An Exposition, First Year, 1828–1829* (Boston: Beacon Press, 1958).

d'Ivray, Jehan. *L'aventure Saint-Simonienne et les femmes* (Paris: Libraire Félix Alcan, 1928).

al-Jabarti, Abd al-Rahman. *Al-Jabarti's Chronicle of the First Seven Months of the French Occupation of Egypt,* edited and translated by S. Moreh (Leiden: E. J. Brill, 1975).

———. *Journal d'un notable du Cairo durant l'éxpedition française, 1798–1801,* translated and annotated by Joseph Cuoq (Paris: Albin Michel, 1979).

James, Lawrence. *The Rise and Fall of the British Empire* (New York: St. Martin's Press, 1994).

Jeanmichel, Lucien. *Arlès-Dufour: un Saint-Simonien à Lyon* (Lyon: Éditions Lyonnaises d'Art et d'Histoire, 1993).

Jenkins, Roy. *Gladstone* (New York: Random House, 1997).

Judd, Denis. *Empire: The British Imperial Experience from 1765 to the Present* (New York: Basic, 1996).

Kabbani, Rana. *Europe's Myths of Orient* (London: Pandora Press, 1986).

Lord Kinross, *Between Two Seas: The Creation of the Suez Canal* (New York: William Morrow, 1969).

———. *The Ottoman Centuries: The Rise and Fall of the Turkish Empire* (New York: Morrow, 1977).

Kissinger, Henry. *Diplomacy* (New York: Simon & Schuster, 1994).

Kyle, Keith. *Suez* (London: Weidenfeld & Nicolson, 1991).

Lambert, Giles. *Auguste Mariette: L'Égypte ancienne sauvée des sables* (Paris: J.C. Lattes, 1997).

Landes, David. *Bankers and Pashas* (New York: Harper & Row, 1958).

Lane, Edward William. *An Account of the Manners and Customs of the Modern Egyptians* (London, 1836).

———. *Description of Egypt* (Cairo: American University in Cairo Press, 2000).

Lapidus, Ira. *A History of Islamic Societies* (Cambridge, Eng.: Cambridge University Press, 1988).

Laurens, Henry. *L'Éxpedition d'Égypte* (Paris: Armand Colin, 1989).

Lawson, Fred. *The Social Origins of Egyptian Expansion During the Muhammad Ali Period* (New York: Columbia University Press, 1992).

Lenczowski, George. *The Middle East in World Affairs: Fourth Edition* (Ithaca: Cornell University Press, 1980).

de Lesseps, Alexandre. *Moi, Ferdinand* (Paris: Olivier Orban, 1986).

de Lesseps, Ferdinand. *Lettres, Journal, et Documents pour servir à l'histoire du canal de Suez,* 5 vols. (Paris: Didier, 1875–1880).

———. *Recollections of Forty Years,* translated by C. B. Pitman (New York: D. Appleton, 1888).

Lewis, Bernard, ed. *A Middle East Mosaic* (New York: Random House, 2000).

———. *The Emergence of Modern Turkey* (New York: Oxford University Press, 1961).

———. *The Middle East and the West* (New York: Harper, 1964).

———. *The Muslim Discovery of Europe* (New York: Norton, 1982).

Mainardi, Patricia. *Arts and Politics of the Second Empire: The Universal Expositions of 1855 and 1867* (New Haven: Yale University Press, 1987).

Mansel, Philip. *Paris Between Empires, 1814–1852* (London: John Murray, 2001).

Manuel, Frank. *The New World of Henri Saint-Simon* (Cambridge: Harvard University Press, 1956).

Marlowe, John. *World Ditch: The Making of the Suez Canal* (New York: Macmillan, 1964).

al-Sayyid Marsot, Afaf Lutfi. *Egypt in the Reign of Muhammad Ali* (Cambridge, Eng.: Cambridge University Press, 1984).

McCullough, David. *The Path Between the Seas* (New York: Simon & Schuster, 1977).

Miquel, Pierre. *Le Second Empire* (Paris: Perrin, 1998).

Mitchell, Timothy. *Colonising Egypt* (New York: Cambridge University Press, 1988).

Montel, Nathalie. *Le Chantier du Canal de Suez* (Paris: Éditions en forma, 1998).

Morsy, Magali, ed. *Les Saint-Simoniens et l'orient* (Édisud: Aix en Provence, 1989).

Nourse, J. E. *The Maritime Canal of Suez* (Washington: Philp & Solomons, 1869).

Owen, Roger. *The Middle East in the World Economy, 1800–1914* (New York: I. B. Tauris, 1993).

Pacha, Nubar. *Mémoires de Nubar Pacha* (Beirut: Libraire de Liban, 1983).

Peltre, Christine. *Orientalism in Art* (New York: Abbeville, 1998).

Peters, Tom. *Building the Nineteenth Century* (Cambridge, Mass.: MIT Press, 1996).

Pincemaille, Christophe. *L'impératrice Eugénie: De Suez à Sedan* (Paris: Payot, 2000).

Plessis, Alain. *De la fête impériale au mur de fédérés 1852–1871* (Paris: Éditions du Seuil, 1979).

Porter, Andrew, ed. *The Oxford History of the British Empire: The Nineteenth Century* (New York: Oxford University Press, 1998).

Porter, Bernard. *The Lion's Share: A Short History of British Imperialism 1850–1983* (London: Longman, 1984).

Price, Roger. *The French Second Empire: An Anatomy of Power* (Cambridge, Eng.: Cambridge University Press, 2001).

Pudney, John. *Suez: De Lesseps' Canal* (New York: Praeger, 1968).

Puryear, Vernon. *France and the Levant from the Bourbon Restoration to the Peace of Kutiah* (Berkeley: University of California Press, 1941).

Rahman, Fazlur. *Islam* (Chicago: University of Chicago Press, 1979).

Raymond, Andre. *Cairo*, translated by Willard Wood (Cambridge, Mass.: Harvard University Press, 2000).

Ridley, Jaspar. *Lord Palmerston* (London: Constable Books, 1970).

Robinson, Ronald, and John Gallagher, *Africa and the Victorians* (London: Macmillan, 1961).

Rodenbeck, Max. *Cairo: The City Victorious* (New York: Knopf, 1998).

Rosenberg, Charles. *The Cholera Years* (Chicago: University of Chicago Press, 1962).

Said, Edward. *Orientalism* (New York: Pantheon, 1977).

———. *Culture and Imperialism* (New York: Knopf, 1993).

Schimmel, Annemarie. *Mystical Dimensions of Islam* (Chapel Hill: University of North Carolina Press, 1975).

Schom, Alan. *Napoleon Bonaparte* (New York: HarperCollins, 1997).

Solé, Robert. *L'Égypte, Passion Française* (Paris: Éditions du Seuil, 1997).

Shaw, Sanford, and Ezel Kural Shaw, *History of the Ottoman Empire and Modern Turkey, vol. II* (Cambridge, Eng.: Cambridge University Press, 1977).

Starkey, Paul, and Janet Starkey, eds. *Travellers in Egypt* (London: Tauris, 1998).

Steele, E. D. *Palmerston and Liberalism, 1855–1865* (Cambridge, Eng.: Cambridge University Press, 1991).

Stiebing, William, *Uncovering the Past: A History of Archeology* (New York: Oxford University Press, 1993).

Taylor, A.J.P. *The Struggle for Mastery in Europe 1848–1918* (New York: Oxford University Press, 1954).

Temperley, Harold. *England and the Near East* (London: Longmans, 1936).

Thompson, J. M. *Napoleon Bonaparte* (Oxford: Blackwell, 1952).

Thomson, David. *England in the Nineteenth Century* (New York: Penguin, 1978).

Thornton, A. P. *The Imperial Idea and Its Enemies* (London: Macmillan, 1985).

Tindall, Gillian. *The Journey of Martin Nadaud* (New York: St. Martin's Press, 1999).

Toledano, Ehud. *State and Society in Mid-Nineteenth-Century Egypt* (New York: Cambridge University Press, 1990).

Tombs, Robert. *France: 1814–1914* (London: Longman, 1996).

Trachtenberg, Marvin. *The Statue of Liberty* (New York: Penguin, 1986).

Trevelyan, George Macauley. *British History in the Nineteenth Century and After* (New York: David McKay, 1937).

Trimingham, J. S. *The Sufi Orders in Islam* (Oxford: Clarendon Press, 1971).

Tugay, Emine Foat. *Three Centuries: Family Chronicles of Turkey and Egypt* (London: Oxford University Press, 1963).

Vatikiotis, P. J. *The History of Egypt*, 3rd ed. (Baltimore: Johns Hopkins University Press, 1985).

Wall, Geoffrey. *Flaubert* (New York: Farrar, Straus & Giroux, 2002).

Webster, Charles. *The Foreign Policy of Lord Palmerston* (London: Bell & Sons, 1951).

Williams, Roger. *The World of Napoleon III, 1851–1870* (New York: Free Press, 1957).

Wilson, Arnold. *The Suez Canal* (New York: Arno Press, 1977).

Yeazell, Ruth Bernard. *Harems of the Mind: Passages of Western Art and Literature* (New Haven: Yale University Press, 2000).

Yousseff, Ahmed. *La Fascination de L'Égypte* (Paris: L'Harmattan, 1998).

Zeldin, Theodore. *France 1848–1945: Politics and Anger* (London: Oxford University Press, 1973).

Zola, Emile. *Nana*, translated by George Holden (New York: Penguin, 1972).

# Acknowledgments

This book is the culmination of many years of study. I have benefited enormously from the wisdom of a number of teachers who have shared their knowledge and urged me to explore the origins of the modern Middle East. At Columbia, as an undergraduate, I was blessed to find myself in an introductory course on the Middle East taught by Richard Bulliet, and his eclectic humor and profound insights have never been far from my mind. Later, at Oxford, Derek Hopwood and Roger Owen prodded me to think more critically about the ebb and flow of Middle Eastern history and dispatched me to Cairo to see for myself.

In researching *Parting the Desert,* I was fortunate to have Yael Schacher as a research assistant. She delved into the tomes and journals, and found sources I might not have stumbled upon without her. She has since begun her own journey toward a Ph.D., which I'm sure she'll complete with the same creative efficiency. I was also aided by the skilled guidance of the staff at the Centre des Archives du Monde du Travail in Roubaix, France, who helped me navigate the records of the Suez Canal Company.

Several scholars agreed to assess the manuscript. Patrice Higonnet of Harvard persuaded me to take a more nuanced aproach toward the Second Empire and Napoleon III, while Peter Marsh of the University of Birmingham encouraged me to reconsider my presentation of Palmerston and mid-nineteenth-century British politics. Finally, Roger Owen meticulously went over the whole text with his usual exacting standards. I have benefited immeasurably from their critiques, and I thank them all for their time and candor. I also had the good fortune to spend some time with Alexandre de Lesseps, who shared his family lore and made sure I was well versed in Ferdinand.

As a friend and an agent, John Hawkins has been an incredible ally and a constant source of support, and at his office, Matthew Miele and Moses Cardona have expertly dealt with details great and small. At Knopf, my editor, Ash Green, has once again demonstrated his pitch-perfect ear for prose that works and prose that doesn't, and his friendship has meant everything. Luba Ostashevsky has managed to keep the production process moving forward, and I thank her for her seemingly endless patience. Paul Bogaards, Kathryn Zuckerman, Nicholas Latimer, and Sheila O'Shea remain the best in the business, and, of course, thanks to Sonny Mehta.

Finally, there are two people who in very different ways made this entire journey possible. Writing this particular book was something of a labor, and I know for a certainty that without my wife, Nicole, by my side, urging me, encouraging me, and succoring me, I would not have had the strength. I hope she knows how vital that was and how vital to me she is. And to Albert Hourani, who taught me to appreciate the melody of the modern Middle East, I hope that this book honors his memory.

# Index

## A NOTE ABOUT THE AUTHOR

Zachary Karabell was educated at Columbia; Oxford, where he received a degree in Modern Middle Eastern Studies; and Harvard, where he received his Ph.D. in 1996. He has taught at Harvard, the University of Massachusetts at Boston, and Dartmouth. He is the author of several books, including *The Last Campaign: How Harry Truman Won the 1948 Election,* which won the *Chicago Tribune* Heartland prize. His essays and reviews have appeared in various publications, such as the *New York Times,* the *Los Angeles Times, Foreign Policy,* and *Newsweek.* He lives in New York City.

## A NOTE ON THE TYPE

This book was set in Fairfield, the first typeface from the hand of the distinguished American artist and engraver Rudolph Ruzicka (1883–1978). In its structure Fairfield displays the sober and sane qualities of the master craftsman whose talent has long been dedicated to clarity. It is this trait that accounts for the trim grace and vigor, the spirited design and sensitive balance, of this original typeface.

Rudolph Ruzicka was born in Bohemia and came to America in 1894. He set up his own shop, devoted to wood engraving and printing, in New York in 1913 after a varied career working as a wood engraver, in photo-engraving and banknote printing plants, and as an art director and free-lance artist. He designed and illustrated many books, and was the creator of a considerable list of individual prints—wood engravings, line engravings on copper, and aquatints.

*Composed by North Market Street Graphics, Lancaster, Pennsylvania*
*Printed and bound by Berryville Graphics, Berryville, Virginia*
*Designed by Robert C. Olsson*